Lecture Notes in Control and Information Sciences

Edited by M. Thoma

For information about Vols. 1—42 please contact your bookseller or Springer-Verlag.

Lecture Notes in Control and Information Sciences

Edited by M. Thoma and A. Wyner

107

Y. T. Tsay, L.-S. Shieh, S. Barnett

Structural Analysis and Design of Multivariable Control Systems

An Algebraic Approach

Springer-Verlag Berlin Heidelberg GmbH

Series Editors
M. Thoma · A. Wyner

Advisory Board
L. D. Davisson · A. G. J. MacFarlane · H. Kwakernaak
J. L. Massey · Ya Z. Tsypkin · A. J. Viterbi

Authors
Yih Tsong Tsay
Leang-San Shieh
Department of Electrical Engineering
University of Houston
Houston, Texas 77004
USA

Stephen Barnett
School of Mathematical Sciences
University of Bradford
West Yorkshire BD7 1DP
United Kingdom

ISBN 978-3-540-18916-9   ISBN 978-3-540-38844-9 (eBook)
DOI 10.1007/978-3-540-38844-9

Preface

Progress in system theory over the last two decades can be broadly categorized into two main streams:

(1) Algebraic System Theory - Study of basic notions and fundamental concepts of both algebra and system theory.

(2) System Analysis and Design Methods - Study of potential design techniques to analyze the characteristics of systems and to design controllers for satisfying various specifications and performance criteria.

Thousands of papers have been published in both areas in the last two decades. Systemic presentations in book form can be found, for example, in [1-5] for the former, in [6-10] for the latter, and in [11-16] for both. From this literature, we find that many elegant theories still cannot be employed to analyze/design the physical systems with ease. In other words, work is still needed to fill the gap between algebraic system theory and practical system analysis/design techniques. This provides the main motivation for our monograph.

The development of our work is based upon state-space representations and matrix fraction descriptions as the mathematical models for physical systems. A unified approach characterizing the dynamics of a system is presented through the formulation of the characteristic λ-matrix (also known as the matrix polynomial) of the system. Applications in pole assignment design, modal control design for multivariable systems, parallel realizations, and cascade realizations of multiport networks are illustrated. A detailed guide to the content of the monograph is provided in the last section of Chapter I.

Acknowledgments

The authors wish to express their gratitude for the fruitful discussions with Dr. Jagdish Chandra, Director of Mathematics Science, US Army Research Office, Dr. Robert E. Yates, Director of the Guidance and Control Directorate, US Army Missile Command, and Dr. Norman P. Coleman, Chief of the Automation and Robotics Group of the Fire Support Laboratory, US Army Armament Research, Development and Engineering Center.

This research monography was financially supported in part by the US Army Research Office under contract DAAL-03-87-K-0001, the US Army Missile Command under contract DAAH-01-85-C-A111, and the NASA-Johnson Space Center under contract NAG9-211.

TABLE OF CONTENTS

CHAPTER I   INTRODUCTION

In this introductory chapter, state-space representations and matrix fraction descriptions of multivariable linear systems are reviewed in Section 1.1. Some basic definitions on λ-matrices, which are the main mathematical tools used in our work, are summarized in Section 1.2, and Section 1.3 gives a guide to the content of the monograph.

1.1 State-Space Representations and Matrix Fraction Descriptions of Multivariable Systems

An m-input, p-output linear time-invariant system σ can be described by state equations as follows:

$$\lambda X(t) = AX(t) + Bu(t) \tag{1.1a}$$

$$y(t) = CX(t) + Du(t) \tag{1.1b}$$

where $X(t) \epsilon X \subseteq C^n$, $y(t) \epsilon Y \subseteq C^p$, $u(t) \epsilon U \subseteq C^m$ are state, output, and input vectors, respectively; X, Y, U are state, output and input spaces of σ, respectively; A,B,C,D are matrices of appropriate dimensions. For continuous-time systems, λ is a differential operator and $t \epsilon R$, while for discrete-time systems, λ is a forward shift operator and $t \epsilon Z$.

Equations (1.1a) and (1.1b) are referred to as the state-space representation of the multivariable system σ. A,B,C, and D can be treated as linear maps:

System map A: $X \to X$

Input map B: $U \to X$

Output map C: $X \to Y$

Forward map D: $U \to Y$ (1.2)

From Eq. (1.2), σ can be described by the following diagram:

$$(1.3)$$

which is not commutative.

The diagram in Eq. (1.3) is useful in studying the structural aspects of the system σ.

From Eq. (1.1), the input-output relationship of the system σ can be represented as

$$y(t) = G(\lambda)u(t) \qquad (1.4a)$$

where

$$G(\lambda) = C(\lambda I_n - A)^{-1}B + D \ \epsilon \, C^{p \times m}(\lambda) \qquad (1.4b)$$

In Eq. (1.4b) $C^{p \times m}(\lambda)$ denotes the set of p×m matrices with elements being rational functions of λ over the complex field C. $G(\lambda)$ is called the transfer function matrix of the system σ. It has been shown in [1,13] that $G(\lambda)$ can be represented as the "ratio" of two matrix polynomials:

$$G(\lambda) = D_\ell^{-1}(\lambda) N_\ell(\lambda) \qquad (1.5a)$$

$$= N_r(\lambda) D_r^{-1}(\lambda) \qquad (1.5b)$$

where $D_\ell(\lambda) \epsilon C^{p \times p}[\lambda]$, $N_\ell(\lambda)$, $N_r(\lambda) \epsilon C^{p \times m}[\lambda]$, $D_r(\lambda) \epsilon C^{m \times m}[\lambda]$; $C^{p \times p}[\lambda]$, $C^{p \times m}[\lambda]$ and $C^{m \times m}[\lambda]$ are sets of matrix polynomials of λ with coefficients in $C^{p \times p}$, $C^{p \times m}$, and $C^{m \times m}$, respectively. Combining Eqs. (1.4) and (1.5), yields

$$y(t) = D_\ell^{-1}(\lambda)N_\ell(\lambda)u(t) \qquad (1.6a)$$

$$= N_r(\lambda)D_r^{-1}(\lambda)u(t) \qquad (1.6b)$$

Equations (1.6a) and (1.6b) are referred to as <u>left matrix fraction descriptions</u> (<u>LMFD</u>) and <u>right matrix fraction descriptions (RMFD)</u> of the system σ, respectively.

Let $T \in C^{n \times n}$ be a nonsingular matrix, and from Eq. (1.1) define

$$\hat{A} = TAT^{-1}, \ \hat{B} = TB, \ \hat{C} = CT^{-1}, \ \hat{D} = D \qquad (1.7a)$$

and

$$\hat{X}(t) = TX(t) \qquad (1.7b)$$

Then the state equations for the system $\hat{\sigma}$ are as follows:

$$\lambda\hat{X}(t) = \hat{A}\hat{X}(t)+\hat{B}u(t) \qquad (1.8a)$$

$$y(t) = \hat{C}\hat{X}(t)+\hat{D}u(t) \qquad (1.8b)$$

For the same set of inputs $u(t)$, σ in Eq. (1.1) and $\hat{\sigma}$ in Eq. (1.8) will generate the same set of outputs $y(t)$ for $t \geq 0$ if $\hat{X}(0) = TX(0)$. The difference between the state vectors $X(t)$ and $\hat{X}(t)$ in the system σ and $\hat{\sigma}$, respectively, is therefore not apparent if only the input-output relationships are considered. Thus, we say that σ and $\hat{\sigma}$ are <u>equivalent systems</u>. Formally, we have the following definition:

<u>Definition 1.1</u>   The system in Eq. (1.1) and the system in Eq. (1.8) are equivalent if and only if the states are related by:

$$\hat{X}(t) = TX(t)$$

We will call this equivalence relation <u>similarity equivalence</u> (SE). □

Let $U_{\ell}(\lambda) \epsilon C^{p \times p}[\lambda]$, and det $U_{\ell}(\lambda) = K_{\ell}$ which is a nonzero constant (i.e. $U_{\ell}(\lambda)$ is <u>unimodular</u>). Define

$$\hat{D}_{\ell}(\lambda) = U_{\ell}(\lambda) D_{\ell}(\lambda) \qquad\qquad (1.9a)$$

$$\hat{N}_{\ell}(\lambda) = U_{\ell}(\lambda) N_{\ell}(\lambda) \qquad\qquad (1.9b)$$

and

$$\hat{G}(\lambda) = \hat{D}_{\ell}^{-1}(\lambda) \hat{N}_{\ell}(\lambda) \qquad\qquad (1.9c)$$

which is an LMFD of a system $\hat{\sigma}$:

$$y(t) = \hat{G}(\lambda) u(t) \qquad\qquad (1.10)$$

From Eqs. (1.6a) and (1.10), σ in Eq. (1.6a) and $\hat{\sigma}$ in Eq. (1.10) will generate the same set of y(t) for t≥0 if the same set of u(t) is used as inputs, and σ and $\hat{\sigma}$ both have the same set of initial conditions y(t), t<0. Thus, we say that σ and $\hat{\sigma}$ are equivalent systems. Similar reasoning can be applied for RMFDs. We reach the following definitions:

<u>Definition 1.2</u>   Two systems with LMFDs $G(\lambda) = D_{\ell}^{-1}(\lambda) N_{\ell}(\lambda)$ and $\hat{G}(\lambda) = \hat{D}_{\ell}^{-1}(\lambda) \hat{N}_{\ell}(\lambda)$ are <u>equivalent</u> if and only if

$$\hat{D}_{\ell}(\lambda) = U_{\ell}(\lambda) D_{\ell}(\lambda)$$

and

$$\hat{N}_{\ell}(\lambda) = U_{\ell}(\lambda) D_{\ell}(\lambda)$$

where $U_\ell(\lambda)$ is unimodular. Similarly, two systems with RMFDs $G(\lambda) = N_r(\lambda)D_r^{-1}(\lambda)$ and $\hat{G}(\lambda) = \hat{N}_r(\lambda)\hat{D}_r^{-1}(\lambda)$ are equivalent if and only if

$$\hat{D}_r(\lambda) = D_r(\lambda)U_r(\lambda)$$

and

$$\hat{N}_r(\lambda) = N_r(\lambda)U_r(\lambda)$$

where $U_r(\lambda)$ is unimodular. We will call this kind of equivalence relations <u>unimodular equivalence</u> (UE). □

It can easily be verified that both SE and UE satisfy the basic properties of an equivalence relation: transitivity, symmetry and reflexivity [13]. Since a system can be represented via state-space equations or matrix fraction descriptions, we have:

<u>Lemma 1.1</u>   Denote SE or UE by $\sim$; σ_x, σ_y and σ_z are systems. Then, we have

(1) Transitivity: $\sigma_x \sim \sigma_y$ and $\sigma_y \sim \sigma_z$ implies $\sigma_x \sim \sigma_z$.

(2) Symmetry: $\sigma_x \sim \sigma_y$ implies $\sigma_y \sim \sigma_x$.

(3) Reflexivity: $\sigma_x \sim \sigma_x$.   ■

From the idea of equivalent systems, both state-space representations and matrix fraction descriptions of multivariable systems are non-unique. In Chapter II, we will develop canonical forms, which are unique for a given system, for both state-space representations and matrix fraction descriptions.

1.2 Fundamental Properties of λ-Matrices

Since the MFD representations of a MIMO (multi-input, multi-output) system involve the ratio of two λ-matrices, and the results presented in the following chapters are closely related to λ-matrices, it is appropriate to review some definitions of λ-matrices in this section. Further details and properties can be found, for example, in [2] and [3]. Specifically, we can define λ-matrices as follows [17,18]. Let F be an arbitrary field, and $F[\lambda]$ be the ring of polynomials over the field F. A λ-matrix, denoted by $A(\lambda)\epsilon F^{p\times m}[\lambda]$ is a $p\times m$ matrix whose elements are in $F[\lambda]$. Let $A_{ij}(\lambda)$ be the (i,j)th element of $A(\lambda)$, then

$$A(\lambda) = (A_{ij}(\lambda)), \quad 1\leq i\leq p, \quad 1\leq j\leq m \qquad (1.11a)$$

and

$$A_{ij}(\lambda) \triangleq \sum_{k=0}^{k_{ij}} a_{ijk}\lambda^{k_{ij}-k}, \quad a_{ijk}\epsilon F \qquad (1.11b)$$

where k_{ij} is the degree of the polynomial $A_{ij}(\lambda)$.

Let $r = Max(k_{ij}, \ 1\leq i\leq p, \ 1\leq j\leq m)$, then $A(\lambda)$ can be written as

$$A(\lambda) = \sum_{k=0}^{r} A_k\lambda^{r-k} \qquad (1.12a)$$

where $A_k\epsilon F^{p\times m}$, and the (i,j)th element of A_k is given by

$$(A_k)_{ij} = \begin{cases} a_{ijk} & \text{if } k\leq k_{ij} \\ 0 & \text{otherwise} \end{cases} \qquad (1.12b)$$

A λ-matrix $A(\lambda)\epsilon F^{p\times p}[\lambda]$ is said to be <u>nonsingular</u> if $det(A(\lambda)) \neq 0$, and <u>regular</u> if the matrix coefficient A_0 of the highest degree term (referred to Eq. (1.12a)) is nonsingular. A regular λ-matrix is <u>monic</u> if A_0 is an identity matrix.

Let $A(\lambda)$ be given by Eq. (1.11), and define

$$\nu_i = Max(k_{ij}, \ 1\leq j\leq m), \quad 1\leq i\leq p \qquad (1.13)$$

Then, ν_i, denoted by $\nu_i = \partial_{r_i}(A(\lambda))$, is the <u>row degree</u> [12] of the ith row of $A(\lambda)$. Similarly

$$\kappa_j = Max(k_{ij}, \ 1 \le i \le p), \ 1 \le j \le m$$

denoted by $\kappa_j = \partial_{c_j}(A(\lambda))$, is the <u>column degree</u> [12] of the jth column of $A(\lambda)$. Define

$$A_{hr} = ((A_{hr})_{ij}), \ 1 \le i \le p, \ 1 \le j \le m \tag{1.15a}$$

where

$$(A_{hr})_{ij} = \begin{cases} a_{ij\nu_i} & \text{if } k_{ij} = \nu_i \\ 0 & \text{if } k_{ij} < \nu_i \end{cases} \tag{1.15b}$$

Then A_{hr} is called the <u>leading row matrix</u> of $A(\lambda)$. $A(\lambda)$ called is a <u>row-reduced</u> λ-matrix if p=m and A_{hr} is nonsingular [12]. Similarly, defining

$$A_{hc} = ((A_{hc})_{ij}), \ 1 \le i \le p, \ 1 \le j \le m \tag{1.16a}$$

where

$$(A_{hc})_{ij} = \begin{cases} a_{ij\kappa_j} & \text{if } k_{ij} = \kappa_j \\ 0 & \text{if } k_{ij} < \kappa_j \end{cases} \tag{1.16b}$$

then A_{hc} is called the <u>leading column matrix</u> of $A(\lambda)$, and if p=m and A_{hc} is nonsingular, $A(\lambda)$ is a <u>column-reduced</u> λ-matrix [12].

To analyze the structure of λ-matrices, it is convenient to transform a general λ-matrix to certain specific forms whose structures can be easily handled. The most commonly used transformations are those of equivalence [12,13].

Definition 1.3   Two λ-matrices $A_1(\lambda)$ and $A_2(\lambda)$ are <u>row equivalent</u>, <u>column equivalent</u>, or <u>equivalent</u>, iff $A_1(\lambda) = U_L(\lambda)A_2(\lambda)$, $A_1(\lambda) = A_2(\lambda)U_R(\lambda)$, or $A_1(\lambda) = U_L(\lambda)A_2(\lambda)U_R(\lambda)$, respectively, where $U_L(\lambda)$ and $U_R(\lambda)$ are unimodular λ-matrices.   □

The equivalence of nonsingular λ-matrices can be stated as follows:

Lemma 1.2   Any nonsingular λ-matrix is row equivalent, column equivalent, or equivalent to a row-reduced, a column-reduced, or a row- and column-reduced λ-matrix.   ■

It is well known that equivalent row-reduced or column-reduced λ-matrices of a given nonsingular λ-matrix [12] are not unique. According to Definition 1.3, a regular λ-matrix is always equivalent to a monic λ-matrix, and the properties and applications of monic λ-matrices have been discussed by many authors [17-32]. We shall extend some known results on monic λ-matrices to row-reduced or column-reduced λ-matrices in the following chapters.

In the analysis and design of multi-input, multi-output (MIMO) systems, MFD representations of the systems are rational matrices over the complex field C. Therefore, we will set $F = C$ in the following chapters whenever λ-matrices are involved.

1.3 Organization of Chapters

The material in this monograph can be regarded as being in two parts: The first part, which includes Chapters II, III and IV, is devoted to exploring the spectral decomposition theory of λ-matrices via the canonical structures of MIMO systems represented in state space equations and MFDs; the second part, which consists of Chapters V and VI, considers applications of the structure theory developed in the first part to the design and decomposition of MIMO systems. Illustrative numerical examples are presented throughout the book.

In Chapter II, the characteristic λ-matrices of multivariable control systems are defined. For a reachable system the characteristic λ-matrix can be constructed from the coefficients of the dependence equations for the column vectors of the reachability test matrix; on the other hand, for an observable system, the left characteristic λ-matrix can be constructed from the coefficients of the dependence equations for the row vectors of the observability test matrix. The controller and observer canonical state-space representations for reachable and observable MIMO systems, respectively, are formally defined. The canonical RMFDs and LMFDs for reachable and observable systems, respectively, are defined, and their properties are discussed based on the canonical controller and observer state-space representations. The characteristic λ-matrices, the canonical state-space forms, and the canonical MFDs are highly dependent on the Kronecker or observability indices of the system. Thus, we also present a numerical method using an orthogonalized projection scheme to compute the Kronecker and observability indices of MIMO system. This numerical algorithm is based on the so-called minimal nice selections.

Spectral analysis of general nonsingular λ-matrices is presented in Chapter III. Firstly, column-reduced and row-reduced canonical λ-matrices are defined; then the equivalent transformations of a nonsingular λ-matrices to a column-reduced or a row-reduced canonical λ-matrix are established. Consequently, the latent roots and latent structures of a general nonsingular λ-matrix can be studied in terms of its equivalent column-reduced or row-reduced canonical λ-matrix. The relationships between the latent structures of nonsingular λ-matrices and the eigenstructures of the system maps in their associated state-space minimal realization quadruples are investigated. As a result, the Jordan chains of nonsingular λ-matrices can be easily found from the input and output maps of their associated Jordan canonical minimal realization quadruples. The matrix roots, formally called solvents, of nonsingular λ-matrices are defined and briefly discussed.

Chapter IV is devoted to developing the theory of divisors and spectral factors of nonsingular λ-matrices. The state-space structures of canonical left and right divisors of nonsingular λ-matrices are extensively investigated via the so-called geometric approaches. Constructive proofs on the existence of the canonical divisors are provided, and some properties of left/right divisors of nonsingular λ-matrices are investigated. Also, the concepts of complete sets of canonical left/right divisors, which are extremely important in the applications to the design and decomposition of MIMO systems, are presented. For completeness, the structures of spectral factorizations used to factor a nonsingular λ-matrix into the product of lower degree canonical λ-matrices are also explored. Finally, computational algorithms for divisors and spectral factors based on block triangularization and block diagonalization of square matrices are discussed. A newly developed matrix sign algorithm is suggested for effective computation of divisors and spectral factors of nonsingular λ-matrices.

The applications of the theory begin in Chapter V, where state-feedback control designs of multivariable systems are studied. Properties of linear state-feedback controls are discussed first. The invariance property of the Kronecker indices of MIMO systems under linear state-feedback controls is an important guide in devising various control schemes. The characteristic λ-matrix and column-reduced λ-matrix assignments for the denominators of the closed-loop MFDs are derived. A study is then made properties of the closed-loop MFDs. For controlling the latent structure of the characteristic λ-matrices in the closed-loop system, we introduce the left/right latent structure assignment. For the purposes of closed-loop decomposition, the divisor assignment and decoupling design, via the notions of divisors, are also presented.

Decomposition theories and their applications to multivariable analysis and design are developed in Chapter VI. Parallel decomposition theory is derived based on the complete sets of left/right divisors of the characteristic λ-matrices. Applications of the parallel decomposition theory to model reduction problem and multiport network synthesis are discussed. The semi-cascade decomposition theory as well as applications to the modal control design of MIMO systems are presented via the notion of spectral factorization of characteristic λ-matrices. Finally, the cascade decomposition theory for MIMO systems is considered, with applications to the cascade realization of multiport networks.

CHAPTER II   CHARACTERISTIC λ-MATRICES AND CANONICAL MATRIX FRACTION DESCRIPTIONS OF MIMO SYSTEMS

It is well-known that the dynamics of single-input, single-output (SISO) systems can be determined from their transfer functions and characteristic polynomials, which are the denominators of the transfer functions. Many system design techniques of SISO systems can actually be interpreted as altering the transfer functions or the characteristic polynomials of the closed loop systems to satisfy some desired dynamic performance criterion. To facilitate the same concepts for MIMO systems, in this chapter we present formal definitions of the characteristic λ-matrices and the canonical MFDs for MIMO systems, which are the counterparts of the characteristic polynomials and the transfer functions for SISO systems, respectively. Properties as well as computational algorithms for the characteristic λ-matrices and the canonical MFDs are given. Relationships between the state space representations and MFDs of MIMO systems are established. The results of this chapter provide the foundations for the spectral analysis of λ-matrices and the decompositions of MIMO systems which are presented in the following chapters.

2.1 Characteristic λ-matrices of Reachable and Observable Systems

An m-input, p-output linear time-invariant system can be described by the state equations as follows:

$$\lambda X(t) = AX(t)+Bu(t) \tag{2.1a}$$

$$y(t) = CX(t)+Du(t) \tag{2.1b}$$

where $X(t) \in C^n$, $y(t) \in C^p$ and $u(t) \in C^m$ are state, output, and input vectors, respectively; A,B,C and D are matrices of appropriate dimensions. For continuous-time systems, λ is a differential operator and $t \in R$, while, for discrete-time systems, λ is a forward shift operator and $t \in Z$.

The system described in Eq. (2.1) is <u>reachable</u> if the reachability test matrix [13]

$$R(A,B) = [B, AB, \ldots, A^{n-1}B] \tag{2.2}$$

is of full rank. The system in Eq. (2.1) is <u>observable</u> if the observability test matrix [13]

$$O(A,C) = [C^T, A^T C^T, \ldots, (A^T)^{n-1} C^T]^T$$
$$= \begin{bmatrix} C \\ CA \\ \vdots \\ CA^{n-1} \end{bmatrix} \tag{2.3}$$

is of full rank.

<u>Definition 2.1</u>   Let $B = [b_1, b_2, \ldots, b_m]$, $b_i \in C^{n \times 1}$. The <u>reachability base matrix</u> is defined by

$$P(A,B) \overset{\Delta}{=} [b_1, Ab_1, \ldots, A^{\kappa_1 - 1} b_1, \ldots, b_m, Ab_m, \ldots, A^{\kappa_m - 1} b_m] \tag{2.4a}$$

where κ_i's are the <u>reachability</u> or <u>Kronecker indices</u> [32-37] of (A,B).   □

<u>Definition 2.2</u>   Let $C = [c_1^T, c_2^T, \ldots, c_p^T]^T$, $c_i \in C^{1 \times n}$. The <u>observability base matrix</u> is defined by

$$Q(A,C) \overset{\Delta}{=} [c_1^T, A^T c_1^T, \ldots, (A^T)^{\nu_1 - 1} c_1^T, \ldots, c_p^T, A^T c_p^T, \ldots, (A^T)^{\nu_p - 1} c_p^T]^T \tag{2.4b}$$

where ν_i's are the <u>observability indices</u> [32-37] of (A,C).   □

Obviously, the columns (rows) of $P(A,B)(Q(A,C))$ are contained in $R(A,B)(O(A,C))$. Thus, $P(A,B)(Q(A,C))$ can be obtained from $R(A,B)(O(A,C))$. The selection of the columns (rows) of $P(A,B)(Q(A,C))$ from $R(A,B)(O(A,C))$ follows the sequence proposed by Popov [35]. It is well known [33-36] that when (A,B) is a reachable pair, $\text{rank}[P(A,B)]=n$ and $\sum_{i=1}^{m} \kappa_i = n$; when (A,C) is an observable pair, then $\text{rank}[Q(A,C)] = n$ and $\sum_{i=1}^{p} \nu_i = n$.

Assume that (A,B) is a reachable pair with reachability indices κ_j, $j=1,2,\ldots,m$. Then $A^{\kappa_j} b_j$ can be uniquely represented as [35-37]

$$A^{\kappa_j} b_j = - \sum_{\substack{i=1 \\ \kappa_j < \kappa_i}}^{j-1} a_{rij(\kappa_j+1)} A^{\kappa_j} b_i - \sum_{i=1}^{m} \sum_{k=1}^{\min[\kappa_i,\kappa_j]} a_{rijk} A^{k-1} b_i , \quad \text{if } \kappa_j > 0,\ 1 \le j \le m$$

(2.5a)

and

$$b_j = - \sum_{\substack{i=1 \\ \kappa_i > 0}}^{j-1} a_{rij1} b_i, \quad \text{if } \kappa_j = 0,\ 1 \le j \le m$$

(2.5b)

Note that the input matrix B may contain dependent vectors. As a result, we can make a general structural analysis of MIMO systems. Similarly, if (A,C) is an observable pair with observability indices ν_i, $i=1,2,\ldots,p$, then $C_i A^{\nu_i}$ can be uniquely represented as

$$C_i A^{\nu_i} = - \sum_{\substack{j=1 \\ \nu_i < \nu_j}}^{i-1} a_{\ell ij(\nu_i+1)} C_j A^{\nu_i} - \sum_{j=1}^{p} \sum_{k=1}^{\min[\nu_i,\nu_j]} a_{\ell ijk} C_j A^{k-1}, \quad \text{if } \nu_i > 0,\ 1 \le i \le p$$

(2.6a)

and

$$C_i = - \sum_{\substack{j=1 \\ \nu_j > 0}}^{i-1} a_{\ell ij1} C_j, \quad \text{if } \nu_i = 0,\ 1 \le i \le p$$

(2.6b)

It has been shown that the Kronecker indices κ_i and the set of parameters $\{a_{rijk}\}$ are invariants of (A,B) under coordinate transformations of states, and the κ_i are also invariant under linear state feedback. Similarly, the observability indices ν_i and the set of parameters $\{a_{\ell ijk}\}$ are invariants of (A,C) under coordinate transformations of states and the ν_i are also invariant for the full state observer construction.

We are now ready to define the characteristic λ-matrices of multi-input multi-output (MIMO) systems [30-32].

Definition 2.3   The <u>right characteristic λ-matrix</u> of a reachable system in Eq. (2.1), or a reachable pair (A,B), is defined by

$$D_r(\lambda) \overset{\Delta}{=} [(D_r(\lambda))_{ij}] \in C^{m \times m}[\lambda], \quad 1 \leq i \leq m, \ 1 \leq j \leq m \tag{2.7a}$$

where the (i,j)th entries of $D_r(\lambda)$ are defined as

$$a_{rii(\kappa_i+1)} \overset{\Delta}{=} 1 \tag{2.7b}$$

$$(D_r(\lambda))_{ij} \overset{\Delta}{=} d_{rij}(\lambda) = \begin{cases} \displaystyle\sum_{k=0}^{\kappa_{ij}} a_{rij(k+1)} \lambda^k & \text{if } \kappa_i > 0, \text{ or } \kappa_i = 0 \text{ and } i = j \\[2ex] 0 & \text{if } \kappa_i = 0, \ i \neq j \end{cases}$$

$$\kappa_{ij} \overset{\Delta}{=} \begin{cases} \kappa_i & \text{if } i = j \\ \kappa_j & \text{if } \kappa_i > \kappa_j, \ i \neq j \\ \kappa_i - 1 & \text{otherwise} \end{cases} \tag{2.7c} \tag{2.7d}$$

An illustrative example is given in Section 2.4.

Definition 2.4   The <u>left characteristic λ matrix</u> of an observable system in Eq. (2.1), or an observable pair (A,C), is defined by

$$D_{\ell}(\lambda) \triangleq [(D_{\ell}(\lambda))_{ij}] \in C^{p \times p}[\lambda], \quad 1 \leq i \leq p, \quad 1 \leq j \leq p \qquad (2.8a)$$

where the (i,j)th entries of $D_{\ell}(\lambda)$ are defined as

$$a_{\ell ii(\nu_i+1)} \triangleq 1 \qquad (2.8b)$$

$$(D_{\ell}(\lambda))_{ij} \triangleq d_{\ell ij}(\lambda) = \begin{cases} \sum_{k=0}^{\nu_{ij}} a_{\ell ij(k+1)}\lambda^k & \text{if } \nu_j>0, \text{ or } \nu_j=0 \text{ and } i=j \\ 0 & \text{if } \nu_j=0 \text{ and } i \neq j \end{cases} \qquad (2.8c)$$

$$\nu_{ij} \triangleq \begin{cases} \nu_j & \text{if } i=j \\ \nu_i & \text{if } \nu_i<\nu_j, \ i \neq j \\ \nu_j-1 & \text{otherwise} \end{cases} \qquad (2.8d) \qquad \square$$

Since $\{a_{rijk}\}$, $\{a_{\ell ijk}\}$ are invariant under coordinate transformations of the states, the right characteristic λ-matrix, $D_r(\lambda)$, and the left characteristic λ-matrix, $D_{\ell}(\lambda)$, are invariant under the coordinate transformations of states [35-36]. Furthermore, since $\kappa_{ij} \leq \kappa_j$ for $1 \leq i \leq m$, the column degree of the jth column of $D_r(\lambda)$ is κ_j. Define the <u>leading column matrix</u> as

$$D_{rh} \triangleq [(D_{rh})_{ij}], \quad 1 \leq i \leq m, \quad 1 \leq j \leq m \qquad (2.9a)$$

$$(D_{rh})_{ij} \triangleq h_{rij} = \begin{cases} a_{rij(\kappa_j+1)} & \text{for } \kappa_j \leq \kappa_i \\ 0 & \text{otherwise} \end{cases} \qquad (2.9b)$$

D_{rh} is constructed using the coefficients of the highest powers of λ in each column. Since $a_{rii(\kappa_i+1)} \equiv 1$, and $a_{rij(\kappa_i+1)} = 0$ for $i>j$, D_{rh} is an upper triangular matrix with diagonal elements all 1's, so $\det(D_{rh}) = 1$. Thus, $D_r(\lambda)$ is a column reduced λ-matrix and the degree of $\det(D_r(\lambda))$ is $\sum_{j=1}^{m} \kappa_j = n$.

Similarly, when $\nu_{ij} \le \nu_i$ for $1 \le j \le p$, the row degree of the ith row of $D_\ell(\lambda)$ is ν_i. Define the _leading row matrix_

$$D_{\ell h} = [(D_{\ell h})_{ij}], \quad 1 \le i \le p, \ 1 \le j \le p \tag{2.10a}$$

$$(D_{\ell h})_{ij} \triangleq h_{\ell ij} = \begin{cases} a_{\ell ij(\nu_i+1)} & \text{for } \nu_i \le \nu_j \\ 0, & \text{otherwise} \end{cases} \tag{2.10b}$$

$D_{\ell h}$ is constructed by the coefficients of the highest powers of λ in each row. Since $a_{\ell ii(\nu_i+1)} \equiv 1$, and $a_{\ell ij(\nu_i+1)} = 0$ for $i < j$, $D_{\ell h}$ is a lower triangular matrix with diagonal elements all 1's, so $\det(D_{\ell h}) = 1$. Thus, $D_\ell(\lambda)$ is a row reduced λ-matrix and the degree of $\det(D_\ell(\lambda))$ is $\sum_{i=1}^{p} \nu_i = n$.

From the definition of the right characteristic λ-matrix in Eq. (2.7) and the properties of D_{rh} in Eq. (2.9), we have the following results:

Proposition 2.1 The right characteristic λ-matrix $D_r(\lambda)$ in Eq. (2.7) has the following properties.

    (a) $d_{rii}(\lambda) \ne 0$, monic

    (b) $\deg d_{rii}(\lambda) > \deg d_{rij}(\lambda), \ i \ne j$

    (c) $\deg d_{rii}(\lambda) > \deg d_{rji}(\lambda), \ j > i$

    (d) $\deg d_{rii}(\lambda) \ge \deg d_{rji}(\lambda), \ j < i$

where $d_{rij}(\lambda)$ is the (i,j)th entry of $D_r(\lambda)$.   □

Similarly, from Eqs. (2.8) and (2.10), we have:

Proposition 2.2 The left characteristic λ-matrix $D_\ell(\lambda)$ in Eq. (2.8) has the following properties.

(a) $d_{\ell i i}(\lambda) \neq 0$, monic

(b) deg $d_{\ell i i}(\lambda) >$ deg $d_{\ell j i}(\lambda)$, i≠j

(c) deg $d_{\ell i i}(\lambda) >$ deg $d_{\ell i j}(\lambda)$, j>i

(d) deg $d_{\ell i i}(\lambda) \geq$ deg $d_{\ell i j}(\lambda)$, j<i

where $d_{\ell i j}(\lambda)$ is the (i,j)th entry of $D_{\ell}(\lambda)$. □

2.2 Canonical Matrix Fraction Descriptions (MFDs) of MIMO Systems

As we have mentioned in Section 1.1, the MFD representations of MIMO systems are nonunique. In order to specify the "standard" MFDs for an MIMO system, we shall define the canonical right MFD for the reachable systems and the canonical left MFD for the observable systems in this section. The relationships between the left/right characteristic λ–matrices and the left/right MFDs of MIMO systems are also discussed.

To find the RMFD of the reachable system in Eq. (2.1), it is convenient to transform the state equations in Eq. (2.1) into the canonical controller form [32–34] using the following similarity transformation:

$$X_c(t) = T_c X(t) \tag{2.11a}$$

$$T_c = [P_1^T, \ldots, (P_1 A^{\kappa_1 - 1})^T, \ldots, P_m^T, \ldots, (P_m A^{\kappa_m - 1})^T]^T \tag{2.11b}$$

where

$$P_i = \text{the } \sigma_i\text{th row of } P^{-1}(A,B), \ \kappa_i > 0 \tag{2.11c}$$

and

$$\sigma_i \triangleq \sum_{j=1}^{i} \kappa_j \tag{2.11d}$$

The state equations in the canonical controller form are:

$$\lambda X_c(t) = A_c X_c(t) + B_c u(t) \qquad\qquad (2.12a)$$

$$y(t) = C_c X_c(t) + D_c u(t) \qquad\qquad (2.12b)$$

where

$$A_c \overset{\Delta}{=} T_c A T_c^{-1} \overset{\Delta}{=} [(A_c)_{ij}], \quad 1 \le i \le m, \ 1 \le j \le m, \ \kappa_i \kappa_j > 0 \qquad\qquad (2.13a)$$

$$(A_c)_{ii} \overset{\Delta}{=} \begin{bmatrix} 0 & \vdots & I_{\kappa_i - 1} \\ \cdots & \vdots & \cdots \\ & A_{cii} & \end{bmatrix} \in C^{\kappa_i \times \kappa_i}; \quad A_{cii} = [-\bar{a}_{rii1}, \ldots, -\bar{a}_{rii\kappa_i}] \qquad (2.13b)$$

$$(A_c)_{ij} \overset{\Delta}{=} \begin{bmatrix} 0 \\ \cdots \\ A_{cij} \end{bmatrix} \in C^{\kappa_i \times \kappa_j}; \quad A_{cij} = \begin{cases} [-\bar{a}_{rij1}, \ldots, -\bar{a}_{rij\kappa_j}], & \kappa_i \le \kappa_j \\[2mm] [-\bar{a}_{rij1}, \ldots, -\bar{a}_{rij\kappa_i}, 0, \ldots, 0], & \kappa_i < \kappa_j \end{cases} \quad i \ne j$$

$$\qquad\qquad (2.13c)$$

and

$$\begin{bmatrix} A_{c11}, \ldots, A_{c1m} \\ \cdot \quad \cdots \quad \cdot \\ A_{cm1}, \ldots, A_{cmn} \end{bmatrix} = -D_{rh}^{-1} \begin{bmatrix} a_{r111}, \ldots, a_{r11\kappa_1}, \ldots, a_{r1m1}, \ldots, a_{r1m\kappa_m} \\ \cdot \quad \cdots \quad \cdot \quad \cdots \quad \cdot \\ a_{rm11}, \ldots, a_{rm1\kappa_1}, \ldots, a_{rmm1}, \ldots, a_{rmm\kappa_m} \end{bmatrix} \qquad (2.13d)$$

$$B_c \overset{\Delta}{=} T_c B = E_{bc} D_{rh}^{-1}; \quad E_{bc} \overset{\Delta}{=} [e_{bc1}, e_{bc2}, \ldots, e_{bcm}] \qquad\qquad (2.13e)$$

$$e_{bci} \overset{\Delta}{=} \begin{cases} e_n^{\sigma_i} & \text{if } \kappa_i > 0 \\[2mm] 0 & \text{if } \kappa_i = 0 \end{cases} \qquad\qquad (2.13f)$$

$$C_c \overset{\Delta}{=} C T_c^{-1} = [C_{c1}^T, C_{c2}^T, \ldots, C_{cp}^T]^T \qquad\qquad (2.13g)$$

$$C_{ci} \triangleq [C_{i11}, \ldots, C_{i1\kappa_i}, \ldots, C_{im1}, \ldots, C_{im\kappa_m}] \tag{2.13h}$$

$$D_c \triangleq D \tag{2.13i}$$

From Eq. (2.12), the input-output relationship of the system can be described as

$$y(t) = [C_c(\lambda I_n - A_c)^{-1}B_c + D_c]u(t) \tag{2.14a}$$

Also, from Eq. (2.13), we have

$$(\lambda I_n - A_c)^{-1}B_c = \psi_r(\lambda)[D_{rh}\delta_r(\lambda)]^{-1} \tag{2.14b}$$

where

$$\psi_r(\lambda) \triangleq [\psi_{r1}(\lambda), \ldots, \psi_{rm}(\lambda)] \tag{2.14c}$$

and

$$\psi_{ri}(\lambda) \triangleq \begin{cases} [0_{1\times\sigma_{i-1}}, 1, \lambda, \ldots, \lambda^{\kappa_i-1}, 0_{1\times(n-\sigma_i)}]^T, & \kappa_i > 0, i = 1, \ldots, m \\ 0_{n\times1}, & \kappa_i = 0 \end{cases} \tag{2.14d}$$

$$\delta_r(\lambda) \triangleq \delta_{rh}(\lambda) - A_r\psi_r(\lambda) \tag{2.14e}$$

$$\delta_{rh}(\lambda) \triangleq \text{diag}[\lambda^{\kappa_i}, i = 1, \ldots, m] \tag{2.14f}$$

$$A_r \triangleq [A_{r1}^T, A_{r2}^T, \ldots, A_{rm}^T]^T \epsilon C^{m\times n} \tag{2.14g}$$

$$A_{ri} \triangleq \begin{cases} \sigma_i\text{th row of } A_c & \text{for } \kappa_i > 0 \\ 0_{1\times n} & \text{for } \kappa_i = 0 \end{cases} \tag{2.14h}$$

From Eq. (2.14a), we have

$$y(t) = [N_r(\lambda)D_r^{-1}(\lambda)+D_c]u(t) \tag{2.15a}$$

where

$$D_r(\lambda) \triangleq D_{rh}\delta_r(\lambda) \tag{2.15b}$$

$$N_r(\lambda) \triangleq C_c\psi_r(\lambda) \tag{2.15c}$$

Thus, the RMFD representation of the system in Eq. (2.1) becomes

$$G(\lambda) = N_r(\lambda)D_r^{-1}(\lambda)+D_c = \hat{N}_r(\lambda)D_r^{-1}(\lambda); \hat{N}_r(\lambda) \triangleq N_r(\lambda)+D_cD_r(\lambda) \tag{2.16}$$

$G(\lambda)$ in Eq. (2.16) is referred to the canonical RMFD of the system in Eq. (2.1), and $D_r(\lambda)$ is the right characteristic λ-matrix of the RMFD. It is well known that, for a reachable system, $D_r(\lambda)$ and $N_r(\lambda)$ are right coprime. From Eqs. (2.14c) and (2.15c), any column of $N_r(\lambda)$ corresponding to $\kappa_i = 0$ is a zero column.

If the system in Eq. (2.1) is observable, the state equations can be transformed into the observer canonical [32-34] form by the similarity transformation as follows:

$$x_0(t) = T_0^{-1}x(t) \tag{2.17a}$$

$$T_0 = [q_1,\ldots,A^{\nu_1-1}q_1,\ldots,q_p,\ldots,A^{\nu_p-1}q_p] \tag{2.17b}$$

where

$$q_i = \text{the } \tau_i\text{th column of } \mathcal{Q}^{-1}(A,C), \ \nu_i > 0 \tag{2.17c}$$

and

$$\tau_i \triangleq \sum_{j=1}^{i} \nu_j \tag{2.17d}$$

The state equations in the observable canonical form are as follows:

$$\lambda X_0(t) = A_0 X_0(t) + B_0 u(t) \tag{2.18a}$$

$$y(t) = C_0 X_0(t) + D_0 u(t) \tag{2.18b}$$

where

$$A_0 \triangleq T_0^{-1} A T_0 = [(A_0)_{ij}], \quad 1 \le i \le p, \ 1 \le j \le p, \ \nu_i \nu_j > 0 \tag{2.19a}$$

$$(A_0)_{ii} \triangleq \begin{bmatrix} 0 & \vdots \\ \cdots & A_{0ii} \\ I_{\nu_i-1} & \vdots \end{bmatrix} \in C^{\nu_i \times \nu_i}, \quad A_{0ii} = [-\bar{a}_{0ii1}, \ldots, -\bar{a}_{0ii\nu_i}]^T \tag{2.19b}$$

$$(A_0)_{ij} \triangleq [0 \ A_{0ij}] \in C^{\nu_i \times \nu_j}; \quad A_{0ij} = \begin{cases} [-\bar{a}_{0ij1}, \ldots, -\bar{a}_{0ij\nu_j}]^T, & \text{if } \nu_i \ge \nu_j, \ i \ne j \\ [-\bar{a}_{0ij1}, \ldots, -\bar{a}_{0ij\nu_i}, 0, \ldots, 0]^T & \text{if } \nu_i < \nu_j, i \ne j \end{cases} \tag{2.19c}$$

and

$$\begin{bmatrix} A_{011}, \ldots, A_{01p} \\ \cdot & \cdots & \cdot \\ A_{0p1}, \ldots, A_{0pp} \end{bmatrix} = - \begin{bmatrix} a_{\ell 111}, \ldots, a_{\ell 11\nu_1}, \ldots, a_{\ell p11}, \ldots, a_{\ell p1\nu_p} \\ \cdot & \cdots & \cdot & \cdots & \cdot & \cdots & \cdot \\ a_{\ell 1p1}, \ldots, a_{\ell 1p\nu_1}, \ldots, a_{\ell pp1}, \ldots, a_{\ell pp\nu_p} \end{bmatrix} \tag{2.19d}$$

$$B_0 \triangleq T_0^{-1} B = [B_{01}, B_{02}, \ldots, B_{0m}] \tag{2.19e}$$

$$B_{0j} \triangleq [b_{1j1}, \ldots, b_{1j\nu_i}, \ldots, b_{pj1}, \ldots, b_{pj\nu_p}]^T \tag{2.19f}$$

$$C_0 \triangleq C T_0 = D_{\ell h}^{-1} E_{c0}; \quad E_{c0} \triangleq [e_{c01}^T, \ldots, e_{c0p}^T]^T \tag{2.19g}$$

$$e_{c0i}^T \triangleq \begin{cases} e_n^{\tau_i}, & \text{if } \nu_i > 0 \\ 0, & \text{if } \nu_i = 0 \end{cases} \tag{2.19h}$$

$$D_0 \overset{\Delta}{=} D \tag{2.19i}$$

From Eq. (2.18), the input-output relationships of the system can be expressed as

$$y(t) = [C_0(\lambda I_n - A_0)^{-1}B_0 + D_0]u(t) \tag{2.20a}$$

From Eq. (2.19), we have

$$C_0(\lambda I_n - A_0)^{-1} = [\delta_\ell(\lambda)D_{\ell h}]^{-1}\psi_\ell(\lambda) \tag{2.20b}$$

where

$$\psi_\ell(\lambda) \overset{\Delta}{=} [\psi_{\ell 1}^T(\lambda), \ldots, \psi_{\ell p}^T(\lambda)]^T \tag{2.20c}$$

and

$$\psi_{\ell i}(\lambda) \overset{\Delta}{=} \begin{cases} [0_{1 \times \tau_{i-1}}, 1, \lambda, \ldots, \lambda^{\nu_i - 1}, 0_{1 \times (n - \tau_i)}], & \nu_i > 0 \\ 0_{1 \times n}, & \nu_i = 0 \end{cases} \tag{2.20d}$$

$$\delta_\ell(\lambda) \overset{\Delta}{=} \delta_{\ell h}(\lambda) - \psi_\ell(\lambda)A_\ell \tag{2.20e}$$

$$\delta_{\ell h}(\lambda) \overset{\Delta}{=} \mathrm{diag}[\lambda^{\nu_i}, i = 1, \ldots, p] \tag{2.20f}$$

$$A_\ell \overset{\Delta}{=} [A_{\ell 1}, A_{\ell 2}, \ldots, A_{\ell p}] \in C^{n \times p} \tag{2.20g}$$

$$A_{\ell i} \overset{\Delta}{=} \begin{cases} \tau_i \text{th column of } A_0 \text{ for } \nu_i > 0 \\ 0_{n \times 1} \text{ for } \nu_i = 0 \end{cases} \tag{2.20i}$$

From Eq. (2.20a), we have

$$y(t) = [D_\ell^{-1}(\lambda)N_\ell(\lambda)+D_0]u(t) \tag{2.21a}$$

where

$$D_\ell(\lambda) = \delta_\ell(\lambda)D_{\ell h} \tag{2.21b}$$

$$N_\ell(\lambda) = \psi_\ell(\lambda)B_0 \tag{2.21c}$$

Thus, the LMFD representation of the system in Eq. (2.1) becomes

$$G(\lambda) = D_\ell^{-1}(\lambda)N_\ell(\lambda)+D_0 = D_\ell^{-1}(\lambda)\hat{N}_\ell(\lambda); \hat{N}_\ell(\lambda) \overset{\Delta}{=} N_\ell(\lambda)+D_\ell(\lambda)D_0 \tag{2.22}$$

$G(\lambda)$ in Eq. (2.22) is referred to the <u>canonical LMFD</u> of the system in Eq. (2.1) and $D_\ell(\lambda)$ is the left characteristic λ-matrix of the LMFD. $D_\ell(\lambda)$ and $N_\ell(\lambda)$ are left coprime if Eq. (2.1) is an observable system. Note that any row of $N_\ell(\lambda)$ corresponding to $\nu_i=0$ is a zero row.

From Eqs. (2.13) and (2.14), the minimal realization of $D_r^{-1}(\lambda)$ using a pair (A,B) can be formulated as follows.

<u>Lemma 2.1</u> The quadruple (A,B,C_r,D_r), where

$$C_r \overset{\Delta}{=} \psi_r^T(0)T_c \quad \text{and} \quad D_r \overset{\Delta}{=} (I_m-\psi_r^T(0)\psi_r(0))D_{rh}^{-1}$$

is a minimal realization of $D_r^{-1}(\lambda)$, i.e. $D_r^{-1}(\lambda) = C_r(\lambda I_n-A)^{-1}B+D_r$.

<u>Proof</u>:

From Eqs. (2.13) and (2.14b), we have

$$T_c(\lambda I_n-A)^{-1}B = \psi_r(\lambda)D_r^{-1}(\lambda) \tag{2.23a}$$

Also, from Eq. (2.14d), we have

$$\psi_r^T(0)\psi_r(\lambda) = \psi_r^T(\lambda)\psi_r(0) = I_m - \sum_{\substack{i=1 \\ \kappa_i=0}}^{m} e_m^i(e_m^i)^T = \psi_r^T(0)\psi_r(0) \qquad (2.23b)$$

Thus,

$$\psi_r^T(0)T_c(\lambda I_n-A)^{-1}B = [I_m - \sum_{\substack{i=1 \\ \kappa_i=0}}^{m} e_m^i(e_m^i)^T]D_r^{-1}(\lambda) \qquad (2.23c)$$

Since $D_r(\lambda) = D_{rh}\delta(\lambda)$, we have $D_r^{-1}(\lambda) = \delta^{-1}(\lambda)D_{rh}^{-1}$. When $\kappa_i=0$, from Eq. (2.14e) we have $\{\delta(\lambda)\}_{ii} = 1$, $\{\delta(\lambda)\}_{ij} = 0$ for $i \neq j$, and $\{\delta(\lambda)\}_{qi} = 0$ for $q \neq i$. As a result, when $\kappa_i=0$, we have $\{\delta^{-1}(\lambda)\}_{ii} = 1$, $\{\delta^{-1}(\lambda)\}_{ij} = 0$ for $i \neq j$, and $\{\delta^{-1}(\lambda)\}_{qi} = 0$ for $q \neq i$. Thus

$$\sum_{\substack{i=1 \\ \kappa_i=0}}^{m} e_m^i(e_m^i)^T D_r^{-1}(\lambda) = [\sum_{\substack{i=1 \\ \kappa_i=0}}^{m} e_m^i(e_m^i)^T]D_{rh}^{-1} = [I_m-\psi_r^T(0)\psi_r(0)]D_{rh}^{-1} \qquad (2.23d)$$

Rearranging the terms in Eq. (2.23c) and substituting Eq. (2.23d) into the rearranged equation yields $D_r^{-1}(\lambda) = C_r(\lambda I_n-A)^{-1}B+D_r$.

It can be easily shown that $(A_c, \psi_r^T(0))$, is an observable pair or (A,C_r) is an observable pair. Thus, the results of Lemma 2.1 follow. ∎

From Lemma 2.1 we observe that, if every $\kappa_i>0$ for $1 \leq i \leq m$, then $D_r = 0_m$, both C_r and B are of full rank, and $D_r^{-1}(\lambda)$ is strictly proper. For representing $D_r(\lambda)$ without involving the inverse of (λI_n-A), an alternative representation of $D_r(\lambda)$ is derived as follows.

Lemma 2.2 $D_r(\lambda)$ can be represented as

$$D_r(\lambda) = \hat{C}_r(\lambda I_n - A)\hat{B}(\lambda) + \hat{D}_r$$

where $\hat{B}(\lambda) \triangleq T_c^{-1}\psi_r(\lambda)$; $\hat{C}_r \triangleq D_{rh}E_{bc}^T T_c$; $\hat{D}_r \triangleq D_{rh}[I_m - \psi_r^T(0)\psi_r(0)]$.

Proof:

Using E_{bc} and A_c in Eq. (2.13), $\psi_r(\lambda)$ and A_r in Eq. (2.14), and $\delta_{rh}(\lambda)$ in Eq. (2.14f), it can be easily verified that $A_r = E_{bc}^T A_c$ and $\delta_{rh}(\lambda) = E_{bc}^T\lambda\psi_r(\lambda) + [I_m - \psi_r^T(0)\psi_r(0)]$. Then, from Eq. (10e) we have

$$\delta(\lambda) = E_{bc}^T\lambda\psi_r(\lambda) + [I_m - \psi_r^T(0)\psi_r(0)] - E_{bc}^T A_c\psi_r(\lambda)$$

$$= E_{bc}^T[\lambda I_n - A_c]\psi_r(\lambda) + [I_m - \psi_r^T(0)\psi_r(0)]$$

$$= E_{bc}^T T_c(\lambda I_n - A)T_c^{-1}\psi_r(\lambda) + [I_m - \psi_r^T(0)\psi_r(0)]$$

Thus, $D_r(\lambda) = D_{rh}\delta(\lambda) = \hat{C}_r(\lambda I_n - A)\hat{B}(\lambda) + \hat{D}_r$. ■

In a similar fashion, from Eqs. (2.19) and (2.20), the minimal realization of $D_\ell^{-1}(\lambda)$ using a pair (A,C) can be expressed as follows:

Lemma 2.3 Define $B_\ell \triangleq T_0\psi_\ell^T(0)$ and $D_\ell \triangleq D_{\ell h}^{-1}[I_p - \psi_\ell(0)\psi_\ell^T(0)]$ then, the quadruple (A, B_ℓ, C, D_ℓ) is a minimal realization of $D_\ell^{-1}(\lambda)$, or $D_\ell^{-1}(\lambda) = C(\lambda I_n - A)^{-1}B_\ell + D_\ell$. ■

Also, an alternative representation of $D_\ell(\lambda)$ without involving the inverse of $(\lambda I_n - A)$ is as follows.

Lemma 2.4 $D_\ell(\lambda)$ can be represented as

$$D_\ell(\lambda) = \hat{C}(\lambda)(\lambda I_n - A)\hat{B}_\ell + \hat{D}_\ell$$

where $\hat{C}(\lambda) = \psi_\ell(\lambda)T_0^{-1}$; $\hat{B}_\ell = T_0 E_{c0} D_{\ell h}$; $\hat{D}_\ell = [I_p - \psi_\ell(0)\psi_\ell^T(0)]D_{\ell h}$. ∎

Lemmas 2.3 and 2.4 can be proven following the same reasoning as in Lemmas 2.1 and 2.2, respectively.

2.3 Minimal Nice Selections for Determining Kronecker and Observability Indices

As we have seen, the right characteristic λ-matrix and the canonical right MFD of a reachable system can be determined once the Kronecker indices of the reachable pair (A,B) are known. To find the Kronecker indices of reachable systems, we introduce the notions of nice selections and minimal nice sections for the reachability test matrix $R(A,B)$. An algorithm to determine the Kronecker indices from the minimal nice selection is also presented. The same procedure is equally applicable for determining the observability indices of an observable system.

Definition 2.5 A nice selection γ is an ordered subset of the set $\{k : 1 \le k \le nm\}$ such that

    (1) The cardinality $|\gamma|$ is equal to n.

    (2) if $i+mj \epsilon \gamma$, then $i+m(j-1) \epsilon \gamma$ for $1 \le i \le m$ and $0 \le j \le n-1$. □

A nice selection γ can be represented by a Young diagram [39] of m rows labeled from 1 to m and n columns labeled from 0 to n-1 with a cross at an (i,j) entry iff $i+mj \epsilon \gamma$ for $1 \le i \le m$ and $0 \le j \le n-1$. Due to the property (2) in Definition 2.5, the Young diagram of a nice selection associated with γ either has consecutive crosses or no crosses at all in each row. The Young diagram with m=3, n=6 and a nice selection $\gamma = \{1,2,3,4,6,9\}$ is illustrated as follows:

28

	j→ 0	1	2	3	4	5	
1	X	X					b_1
i→ 2	X						b_2
3	X	X	X				b_3
	I_3	A	A^2	A^3	A^4	A^5	

I_m denotes an mxm identity matrix.

<u>Definition 2.6</u> Let (A,B) be a reachable pair and ℓ_j be the number of crosses in the jth column of the Young diagram of a nice selection γ associated with (A,B). If

$$\sum_{j=0}^{k-1} \ell_j = \text{rank}(B,AB,\ldots,A^{k-1}B), \quad k \leq n \tag{2.24}$$

and $(A^j b_i)$, $\forall (i,j) \in \{(\alpha,\beta): 1 \leq \alpha \leq m, 0 \leq \beta \leq n-1 \text{ and } \alpha+m(\beta-1) \in \gamma\}$ are independent, then γ, defined as γ_M, is a <u>minimal</u> nice selection of the reachable pair (A,B). □

It has been shown [37] that minimal nice selections are unique up to permutations of the rows of the Young diagram. Also, if κ_i denotes the number of crosses at the ith row of the Young diagram of a minimal nice selection γ_M, the set $\kappa = \{\kappa_i, 1 \leq i \leq m\}$ is the Kronecker indices of the reachable pair (A,B) associated with γ_M. Note that κ_i may be zero for some i. The maximum number of minimal nice selections associated with a reachable pair (A,B) is m!, and each minimal nice selection results in one controller form and one associated RMFD. All the controller forms and associated RMFDs are equivalent. To determine the canonical controller form and associated canonical RMFD, we define the input selection sequence as follows.

<u>Definition 2.7</u> The input selection sequence for a multivariable system with m-input, $S \overset{\Delta}{=} \{s_k, 1 \leq k \leq m\}$, is defined as a permutation of the set $\{k, 1 \leq k \leq m\}$. The

natural input selection sequence, S_N, is defined as $S_N \overset{\Delta}{=} \{s_1, s_2, \ldots, s_m\} = \{1, 2, \ldots, m\}$. The minimal nice selection γ_M with the input selection sequence S is determined by examining the independent vectors of the ordered column vectors in the following matrix:

$$\bar{R}(A,B) \overset{\Delta}{=} [b_{s_1}, b_{s_2}, \ldots, b_{s_m}, Ab_{s_1}, Ab_{s_2}, \ldots, Ab_{s_m}, \ldots, A^{n-1}b_{s_1}, A^{n-2}b_{s_2}, \ldots, A^{n-1}b_{s_m}]$$

$$(2.25a)$$

$\square$

If the set κ contains the Kronecker indices of the minimal nice selection γ_M of a reachable pair (A,B) with the natural input selection sequence S_N, then the reachability base matrix of the (A,B) due to Popov [35] is defined as

$$P(A,B) = [b_1, Ab_1, \ldots, A^{\kappa_1-1}b_1, \ldots, b_m, Ab_m, \ldots, A^{\kappa_m-1}b_m] \qquad (2.25b)$$

and the controller form transformation matrix T_c becomes Eq. (2.11b). From Lemma 2.1, the right characteristic λ-matrix can be determined. In addition, the RMFD can be found by using Eqs. (2.14) and (2.15).

The linearly independent vectors in a rectangular or square matrix $R(A,B)$ can often be determined by applying the Sylvester theorem [13] to the square symmetric matrix $R(A,B)R^T(A,B)$, or by calculating the nonzero elements at the diagonal entries of two triangular matrices, decomposed from the square symmetric matrix $R(A,B)R^T(A,B)$ [38]. Here, we propose an orthogonalized projection approach to determine the linearly independent vectors of $R(A,B)$ in Eq. (2.25a) as follows.

Let $V = \{v_i, 1 \le i \le N\}$ be a set of vectors $v_i \in C^{n \times 1}, N \le n$. The orthogonalized projection $P(k) \in C^{m \times m}$ for $V_k \overset{\Delta}{=} \{v_i, 1 \le i \le k\}$, $1 \le k \le N$, can be evaluated recursively as follows:

$$p(0) = I_n \quad \text{(the unit matrix of order n)} \qquad (2.26a)$$

$$P(k) = \begin{cases} P(k-1) - \dfrac{P(k-1)v_k v_k^* P(k-1)}{v_k^* P(k-1)v_k} & \text{for } v_k^* P(k-1)v_k \neq 0 \\[3ex] P(k-1) & \text{for } v_k^* P(k-1)v_k = 0 \end{cases} \qquad (2.26b)$$

where v_k^* denotes the conjugate transpose of v_k.

<u>Lemma 2.5</u> Some important properties of the orthogonalized projection [39] are as follows:

    (1) $P^*(k) = P(k)$ or $P(k)$ is symmetric.

    (2) $P^2(k) = P(k)$ or $P(k)$ is idempotent.

    (3) $P(k)v_i = 0_{n \times 1}$, $\forall\ v_i \in V_k$, $1 \leq i \leq k$.

    (4) $P(k)X = X$, $\forall\ X \perp V_k$, $X \in C^{n \times 1}$, $1 \leq i \leq k$.   ■

From Lemma 2.5, we have the following results:

<u>Lemma 2.6</u> $v_k^* P(k-1)v_k = 0$ iff $v_k \in V_{k-1}$, where $V_{k-1} \overset{\Delta}{=} \text{Span}_c[V_{k-1}]$.

<u>Proof:</u>

    Since $P(k-1)$ is idempotent and symmetric, we have

$$v_k^* P(k-1)v_k = v_k^* P^*(k-1)P(k-1)v_k = \left(P(k-1)v_k\right)^* \left(P(k-1)v_k\right)$$

(a) If $v_k \in V_{k-1}$, then $v_k = \sum_{i=1}^{k-1} a_i v_i$. From property (3) of Lemma 2.5 we have

$$P(k-1)v_k = \sum_{i=1}^{k-1} a_i P(k-1)v_i = 0_{n \times 1}$$

    Thus

$$v_k^* P(k-1)v_k = 0$$

(b) If $v_k^* P(k-1)v_k = 0$, we have $P(k-1)v_k = 0_{n \times 1}$.

Let

$$v_k = \sum_{i=1}^{k-1} a_i v_i + \hat{v}_k, \quad \hat{v}_k \perp_{k-1}$$

Then, from property (4) of Lemma 2.5 we obtain

$$P(k-1)v_k = P(k-1)\hat{v}_k = \hat{v}_k = 0_{n\times 1}$$

Thus, $v_k = \sum_{i=1}^{k-1} a_i v_i$ or $v_k \in _{k-1}$. ∎

Lemma 2.7 Define $W_k \subset V_k$; $W_0 \overset{\Delta}{=} \phi$;

$$W_k \overset{\Delta}{=} \begin{cases} W_{k-1} \cup \{v_k\} & \text{if } v_k^* P(k-1)v_k \neq 0 \text{ for } 1 \leq k \leq N \\ W_{k-1} & \text{if } v_k^* P(k-1)v_k = 0 \quad \text{for } 1 \leq k \leq N \end{cases}$$

Then, W_N is a base for V_N.

Proof:

Lemma 2.7 can be proved directly from Lemma 2.6. ∎

Lemma 2.7 provides a foundation for deriving the recursive algorithm for selecting the independent vectors from a set of vectors. Based on Lemma 2.7, we now derive an algorithm to determine the Kronecker indices of the dynamic system in Eq. (2.1) via the minimal nice selection as follows [40].

32

Algorithm 2.1

```
Given:  (A,B) - A reachable pair
        S    - The input selection sequence
Find:   κ    - The Kronecker indices
Algorithm:
  {Initialization}
  P:=I_n {set orthogonalized projection to identify matrix}
  For i:=1 to m Do
   Begin
     κ_i:=0;{Reset Kronecker indices}
     Flag_i:=True;{Set selection flags true for all inputs}
     v_i:=b_i {Copy b_i to v_i}
   End;
  {processing}
  Repeat
  For i:=1 to m Do
   Begin
   j:=s_i ; {Select A^k b_{s_i}, k≤0}
   If Flag_j then
     Begin
       d:=P*v_j;
       If v_j*d≠0 then
         Begin {Select v_j}
           P:=P-d*d*/v_j*d;{Update P}
           v_j:=A*v_j;{Update v_j for next selection}
           κ_j:=κ_j+1{Update Kronecker index}
         End
       Else
             Flag_j:=false {Reset selection Flag}
     End
   End {For loop}
  Until all Flag_i are false;
```

Note that the * between two variables in the above algorithm is the product notation, and the superscript * designates conjugate transpose. ∎

When the system in Eq. (2.1) is not reachable, the sum of the Kronecker indices κ_i found by using the above algorithm is less than n, i.e. $\sum\limits_{i=1}^{m} \kappa_i < n$. Thus, the above algorithm can also be used to test the reachability of systems in state space form.

As is well-known, reachability and observability of a system are dual concepts. The above algorithm can be modified to find the observability indices ν_i by substituting (A^T, C^T) for (A,B), ν_i for κ_i, and treating S as the output selection sequence which is defined in the same way as the input selection sequence in Definition 2.6.

2.4 Illustrative Examples

Consider the following 3-input 2-output system

$$\lambda X(t) = AX(t) + Bu(t) \tag{2.27a}$$

$$y(t) = CX(t) + Du(t) \tag{2.27b}$$

where

$$A = \begin{bmatrix} -1 & 0 & -2 & 5 & -9 \\ 0 & -1 & 4 & -8 & 14 \\ 0 & 0 & 1 & -4 & 7 \\ 0 & 0 & 0 & -1 & 6 \\ 0 & 0 & 0 & 0 & 2 \end{bmatrix} ; \quad B = \begin{bmatrix} 4 & -3 & -2 \\ -1 & 1 & 1 \\ -1 & 3 & 5 \\ 2 & 1 & 4 \\ 1 & 0 & 1 \end{bmatrix}$$

$$C = \begin{bmatrix} 1 & -1 & 4 & -5 & 11 \\ 0 & 1 & -1 & 4 & -5 \end{bmatrix} ; \quad D = 0_{2 \times 3}$$

Determine various controller forms and associated RMFDs using minimal nice selection γ_M and various input selection sequences S.

(1) If the input selection sequence, S, is selected as S = {1,2,3}, which is the __natural__ input selection sequence S_N, then the __canonical__ controller form and associated __canonical__ RMFD can be determined as follows.

Using the minimal nice selection algorithm, we obtain the Kronecker indices $\kappa_1=2$, $\kappa_2=3$ and $\kappa_3=0$. Note that the input matrix B contains a dependent vector. The corresponding Young diagram becomes

The state equations in the **canonical** controller form of Eq. (2.27) can be determined from Eqs. (2.11) and (2.12) as follows:

$$A_c = \begin{bmatrix} 0 & 1 & 0 & 0 & 0 \\ 2 & -1 & 0 & 0 & 0 \\ \hline 0 & 0 & 0 & 1 & 0 \\ 0 & 0 & 0 & 0 & 1 \\ 0 & 0 & 1 & 1 & -1 \end{bmatrix}; \quad B_c = \begin{bmatrix} 0 & 0 & 0 \\ 1 & 0 & 1 \\ 0 & 0 & 0 \\ 0 & 0 & 0 \\ 0 & 1 & 2 \end{bmatrix} = \begin{bmatrix} 0 & 0 & 0 \\ 1 & 0 & 0 \\ 0 & 0 & 0 \\ 0 & 0 & 0 \\ 0 & 1 & 0 \end{bmatrix}\begin{bmatrix} 1 & 0 & 1 \\ 0 & 1 & 2 \\ 0 & 0 & 1 \end{bmatrix} = E_{bc}D_{rh}^{-1}$$

$$C_c = \begin{bmatrix} -4 & 2 & -2 & 3 & 3 \\ 0 & 3 & 0 & 2 & 2 \end{bmatrix}; \quad D_c = D = 0_{2\times3} \tag{2.28a}$$

The right characteristic λ-matrix $D_r(\lambda)$ can be determined from Eq. (2.15) as

$$D_r(\lambda) = D_{rh}\delta_r(\lambda) = D_{rh}[\delta_{rh}(\lambda)-A_r\psi_r(\lambda)]$$

$$= \begin{bmatrix} \lambda^2-\lambda-2 & 0 & -1 \\ 0 & \lambda^3+\lambda^2-\lambda-1 & -2 \\ 0 & 0 & 1 \end{bmatrix} \tag{2.28b}$$

where

$$D_{rh} = \begin{bmatrix} 1 & 0 & 1 \\ 0 & 1 & 2 \\ 0 & 0 & 1 \end{bmatrix}^{-1} = \begin{bmatrix} 1 & 0 & -1 \\ 0 & 1 & -2 \\ 0 & 0 & 1 \end{bmatrix} \; ; \quad \delta_{rh}(\lambda) = \begin{bmatrix} \lambda^2 & 0 & 0 \\ 0 & \lambda^3 & 0 \\ 0 & 0 & 1 \end{bmatrix}$$

$$A_r = \begin{bmatrix} 2 & 1 & 0 & 0 & 0 \\ 0 & 0 & 1 & 1 & -1 \\ 0 & 0 & 0 & 0 & 0 \end{bmatrix} \; ; \quad \psi_r(\lambda) = \begin{bmatrix} 1 & \lambda & 0 & 0 & 0 \\ 0 & 0 & 1 & \lambda & \lambda^2 \\ 0 & 0 & 0 & 0 & 0 \end{bmatrix}^T$$

Note that D_{rh} is an upper triangular matrix with diagonal entries all 1's, and $D_r(\lambda)$ is a column-reduced λ-matrix which satisfies all the conditions in Proposition 2.1.

The <u>canonical</u> RMFD can be obtained from Eqs. (2.14) and (2.15) as follows:

$$y(t) = N_r(\lambda) D_r^{-1}(\lambda) u(t) \tag{2.28c}$$

where

$$N_r(\lambda) = C\psi_r(\lambda) = \begin{bmatrix} 2\lambda-4 & 3\lambda^2+3\lambda-2 & 0 \\ 3\lambda & 2\lambda^2+2\lambda & 0 \end{bmatrix}$$

(2) If the input selection sequence, S, is selected as $S = \{2,3,1\}$, then the Kronecker indices become $\kappa_1=0$, $\kappa_2=3$, and $\kappa_3=2$ and the Young diagram is

	0	1	2	3	4	
1						b_1
2	X	X	X			b_2
3	X	X				b_3
	I_5	A	A^2	A^3	A^4	

The state equations in a controller form of Eq. (2.27) become

$$
A_c = \begin{bmatrix} 0 & 1 & 0 & 0 & 0 \\ 0 & 0 & 1 & 0 & 0 \\ 1 & 1 & -1 & 0 & 0 \\ \hline 0 & 0 & 0 & 0 & 1 \\ 0 & 0 & 0 & 2 & 1 \end{bmatrix} \; ; \quad
B_c = \begin{bmatrix} 0 & 0 & 0 \\ 0 & 0 & 0 \\ 0 & 1 & 2 \\ \hline 0 & 0 & 0 \\ 1 & 0 & 1 \end{bmatrix} = \begin{bmatrix} 0 & 0 & 0 \\ 0 & 0 & 0 \\ 0 & 1 & 0 \\ \hline 0 & 0 & 0 \\ 0 & 0 & 1 \end{bmatrix} \begin{bmatrix} 1 & 0 & 0 \\ 0 & 1 & 2 \\ 1 & 0 & 1 \end{bmatrix} = E_{bc} D_{rh}^{-1}
$$

where

$$
D_{rh} = \begin{bmatrix} 1 & 0 & 0 \\ 0 & 1 & 2 \\ 1 & 0 & 1 \end{bmatrix}^{-1} = \begin{bmatrix} 1 & 0 & 0 \\ 2 & 1 & -2 \\ -1 & 0 & 1 \end{bmatrix}
$$

$$
C_c = \begin{bmatrix} -2 & 3 & 3 & -4 & 2 \\ 0 & 2 & 2 & 0 & 3 \end{bmatrix} \tag{2.29a}
$$

Note that D_{rh} is not a triangular matrix.

The corresponding RMFD can be obtained from Eqs. (2.14) and (2.15) as follows:

$$
y(t) = N_{r1}(\lambda) D_{r1}^{-1}(\lambda) u(t) \tag{2.29b}
$$

where

$$
D_{r1}(\lambda) = \begin{bmatrix} 1 & 0 & 0 \\ 2 & \lambda^3 + \lambda^2 - \lambda - 1 & -2\lambda^2 + 2\lambda + 4 \\ -1 & 0 & \lambda^2 - \lambda - 2 \end{bmatrix} \; ; \quad
N_{r1}(\lambda) = \begin{bmatrix} 0 & 3\lambda^2 + 3\lambda - 2 & 2\lambda - 4 \\ 0 & 2\lambda^2 + 2\lambda & 3\lambda \end{bmatrix}
$$

Note that $D_{r1}(\lambda)$ is a column-reduced λ-matrix but it does not satisfy the conditions (b) and (c) in Proposition 2.1.

(3) If the input selection sequence, S, is selected as S = {3,1,2}, then the Kronecker indices become $\kappa_1 = 2$, $\kappa_2 = 0$, and $\kappa_3 = 3$, and the Young diagram is

	0	1	2	3	4	
1	X	X				b_1
2						b_2
3	X	X	X			b_3
	I_5	A	A^2	A^3	A^4	

The state equations in a controller form become

$$
A_c = \begin{bmatrix} 0 & 1 & 0 & 0 & 0 \\ 2 & 1 & 3 & 3 & 0 \\ 0 & 0 & 0 & 1 & 0 \\ 0 & 0 & 0 & 0 & 1 \\ 0 & 0 & 1 & 1 & -1 \end{bmatrix} ; \quad B_c = \begin{bmatrix} 0 & 0 & 0 \\ 1 & -5 & 0 \\ 0 & 0 & 0 \\ 0 & 0 & 0 \\ 0 & 5 & 1 \end{bmatrix} = \begin{bmatrix} 0 & 0 & 0 \\ 1 & 0 & 0 \\ 0 & 0 & 0 \\ 0 & 0 & 0 \\ 0 & 0 & 1 \end{bmatrix} \begin{bmatrix} 1 & -5 & 0 \\ 0 & 1 & 0 \\ 0 & 5 & 1 \end{bmatrix} = E_{bc} D_{rh}^{-1}
$$

where

$$
D_{rh} = \begin{bmatrix} 1 & -5 & 0 \\ 0 & 1 & 0 \\ 0 & 5 & 1 \end{bmatrix}^{-1} = \begin{bmatrix} 1 & 5 & 0 \\ 0 & 1 & 0 \\ 0 & -5 & 1 \end{bmatrix}
$$

$$
C_c = \begin{bmatrix} -4 & 2 & -12 & 6 & 8 \\ 0 & 3 & 0 & 10 & 7 \end{bmatrix} \tag{2.30a}
$$

The corresponding RMFD can be described as follows:

$$
y(t) = N_{r2}(\lambda) D_{r2}^{-1}(\lambda) u(t) \tag{2.30b}
$$

where

$$
D_{r2}(\lambda) = \begin{bmatrix} \lambda^2-\lambda-2 & 5 & -3\lambda-3 \\ 0 & 1 & 0 \\ 0 & -5 & \lambda^3+\lambda^2-\lambda-1 \end{bmatrix} ; \quad N_{r2}(\lambda) = \begin{bmatrix} 2\lambda-4 & 0 & 8\lambda^2+6\lambda-12 \\ \\ 3\lambda & 0 & 7\lambda^2+10\lambda \end{bmatrix}
$$

Note that in this case D_{rh} is not a triangular matrix, and $D_{r2}(\lambda)$ is a column-reduced λ-matrix but it does not satisfy the conditions (b) and (c) in Proposition 2.1.

In this chapter, the latent structures of nonsingular λ-matrices are investigated. The column-reduced and the row-reduced canonical λ-matrices are defined as the standard forms of a nonsingular λ-matrix. Equivalence transformations are developed for transforming nonsingular λ-matrices to row-reduced or column-reduced canonical λ-matrices, and for transforming non-canonical MFDs to canonical MFDs.

The existence of the first types of transformations enables us to analyze the latent structures of nonsingular λ-matrices in terms of the latent structures of canonical λ-matrices. Due to the simplicity of the formulation for minimal realizations of the inverse of the canonical λ-matrices in state-space, the latent structures of canonical λ-matrices can easily be investigated from their associated state-space minimal realization quadruples.

After the latent structures of nonsingular λ-matrices are established, the "matrix roots", called solvents, of nonsingular λ-matrices are studied via the latent roots and latent vectors. Solvents of monic λ-matrices have been discussed in many previous works [17-18,22-30]. Our reasons for considering solvents in this chapter are: (1) to extend the idea of solvents for nonsingular λ-matrices; (2) to develop the structure of a special class of divisors, namely monic linear divisors of nonsingular λ-matrices. The general structure of divisors of nonsingular λ-matrices is studied in the next chapter.

3.1 Canonical λ-matrices and Canonical MFDs of MIMO Systems

In this section, we develop a technique to reduce a nonsingular λ-matrix to a row-reduced or column-reduced canonical λ-matrix. The reduction technique is then extended to convert a general MFD to one whose denominator is a row-reduced or column-reduced canonical λ-matrix. The converted MFDs are referred to as canonical MFDs.

<u>Definition 3.1</u> A <u>column-reduced canonical λ-matrix</u>, $D_r(\lambda)$, can be described as

$$D_r(\lambda) = D_{rh}\delta_r(\lambda)$$

where D_{rh} is an upper triangular square matrix with diagonal elements being all 1's. $\delta_r(\lambda)$ is a λ-matrix with the (i,i)th entry, $(\delta_r(\lambda))_{i,i}$, being a monic λ polynomial of degree $\kappa_i = \partial_{ci}(D_r(\lambda))$ and the (i,j)th entry, $(\delta_r(\lambda))_{i,j}$ being a λ polynomial of degree $\leq \min(\kappa_i, \kappa_j)-1$, i≠j.

A <u>row-reduced canonical λ-matrix</u>, $D_\ell(\lambda)$, can be described as

$$D_\ell(\lambda) = \delta_\ell(\lambda)D_{\ell h}$$

where $D_{\ell h}$ is a lower triangular square matrix with diagonal elements being all 1's. $\delta_\ell(\lambda)$ is a λ-matrix with the (i,i)th entry, $(\delta_\ell(\lambda)_{i,i}$ being a monic λ polynomial of degree $\nu_i = \partial_{ri}(D_\ell(\lambda))$ and the (i,j)th entry, $(\delta_\ell(\lambda))_{i,j}$ being a λ polynomial of degree $\leq \min(\nu_i, \nu_j)-1$, i≠j. □

From Definition 3.1, we observe that the left (right) characteristic λ-matrices of MIMO systems discussed in Chapter II are the row (column) reduced <u>canonical</u> λ-matrices. Their salient feature is that the inverse of these λ-matrices can be described by the state-space realizations in canonical <u>observer</u> or <u>controller</u> forms. Therefore, the structural properties of these λ-matrices can be systematically explored in terms of a state-space setting.

As mentioned in Lemma 1.2 of Chapter I, a nonsingular λ-matrix is row equivalent, column equivalent or equivalent to a row-reduced, a column-reduced or a row-and-column-reduced λ-matrix. However, since equivalent row-reduced (column-reduced) λ-matrices of a given nonsingular λ-matrix are not unique, a row-reduced (column-reduced) λ-matrix is not necessarily a row-reduced (column-reduced) canonical λ-matrix as defined in Definition 3.1. We will investigate a constructive way to explore the equivalent transformation of a nonsingular λ-matrix to its row-reduced or column-reduced canonical λ-matrix as follows.

We begin with the state-space minimal realizations of the inverse of the reduced λ-matrices.

Lemma 3.1 Let $A(\lambda) \in C^{m \times m}[\lambda]$ be a column-reduced λ-matrix with the column degrees $\hat{\kappa}_i, i=1,\ldots,m$. Then $A(\lambda)$ can be represented as

$$A(\lambda) = D_{ch}[D_h(\lambda) - A_{cr}\psi_c(\lambda)]$$

where D_{ch} is the leading column matrix of $A(\lambda)$ and $D_h(\lambda) \triangleq \text{diag}[\lambda^{\hat{\kappa}_i}, i=1,\ldots,m]$; $\psi_c(\lambda) \triangleq [\psi_{c1}(\lambda),\psi_{c2}(\lambda),\ldots,\psi_{cm}(\lambda)]$; $\psi_{ci}(\lambda) \triangleq [0_{1 \times \hat{\sigma}_{i-1}},1,\lambda,\ldots,\lambda^{\hat{\kappa}_i-1},0_{1 \times (n-\hat{\sigma}_i)}]^T$ for $\hat{\kappa}_i>0$ and $\psi_{ci}(\lambda) \triangleq 0_{n \times 1}$ for $\hat{\kappa}_i=0$; $\hat{\sigma}_i \triangleq \sum_{j=1}^{i} \hat{\kappa}_j$, $i=1,\ldots,m$; $n \triangleq \hat{\sigma}_m$; $A_{cr} \in C^{m \times n}$ can be determined from D_{ch} and the coefficients of each entry of $A(\lambda)$. ∎

The minimal realization of $A^{-1}(\lambda)$ can be stated below.

Lemma 3.2 $\qquad A^{-1}(\lambda) = \hat{C}_c(\lambda I_n - \hat{A}_c)^{-1}\hat{B}_c + \hat{D}_c$ $\qquad\qquad\qquad$ (3.1a)

where

$; \quad \hat{n} \triangleq n - \hat{\kappa}_1 - \hat{\kappa}_m$

$\qquad\qquad\qquad\qquad\qquad\qquad\qquad\qquad\qquad\qquad\qquad\qquad\qquad\qquad\qquad$ (3.1b)

$(A_{cr})_i \triangleq$ The ith row of A_{cr}, $i=1,\ldots,m$ $\qquad\qquad\qquad\qquad\qquad\qquad$ (3.1c)

$$\hat{B}_c \triangleq \hat{E}_{bc}D_{ch}^{-1}; \quad \hat{E}_{bc} \triangleq [\hat{e}_{bc1},\hat{e}_{bc2},\ldots,\hat{e}_{bcm}]; \quad \hat{e}_{bci} \triangleq \begin{cases} e_n^{\hat{\sigma}_i} & \text{if } \hat{\kappa}_i>0 \\ 0_{n \times 1} & \text{if } \hat{\kappa}_i=0 \end{cases} \qquad (3.1d)$$

$$\hat{C}_c \triangleq \psi_c^T(0) \tag{3.1e}$$

$$\hat{D}_c \triangleq [I_m - \psi_c^T(0)\psi_c(0)]D_{ch}^{-1} \tag{3.1f}$$

Proof:

Lemma 3.2 can be established by direct verification. ∎

The quadruple $(\hat{A}_c, \hat{B}_c, \hat{C}_c, \hat{D}_c)$ is referred to as a <u>controller</u> (<u>not</u> a canonical controller) <u>form realization</u> of $A^{-1}(\lambda)$. In general, we can define $A \triangleq \hat{T}_c^{-1}\hat{A}_c\hat{T}_c$, $B \triangleq \hat{T}_c^{-1}\hat{B}_c$, $C \triangleq \hat{C}_c\hat{T}_c$ and $D = \hat{D}_c$, where $\hat{T}_c$ is a non-singular similarity transformation matrix, so that the inverse of a column-reduced λ-matrix, $A(\lambda)$, can be realized by using a quadruple (A,B,C,D), or

$$A^{-1}(\lambda) = C(\lambda I_n - A)^{-1}B + D \tag{3.2}$$

Let $D_r(\lambda)$ be the right characteristic λ-matrix of (A,B) in Eq. (3.2). Using Lemma 2.1, $D_r^{-1}(\lambda)$ can be realized as follows:

$$D_r^{-1}(\lambda) = C_r(\lambda I_n - A)^{-1}B + D_r . \tag{3.3}$$

The relationship between $A(\lambda)$ in Eq. (3.2) and $D_r(\lambda)$ in Eq. (3.3) is the following:

<u>Theorem 3.1</u> Let $A(\lambda)$ be a column-reduced λ-matrix represented by Eq. (3.2), and let $D_r(\lambda)$ be defined in Eq. (3.3). Then,

$$A(\lambda) = D_r(\lambda)U_r(\lambda) \tag{3.4}$$

where $U_r(\lambda)$ is a unimodular λ-matrix, $U_r(\lambda) = [CT_c^{-1}\psi_r(\lambda)+DD_r(\lambda)]^{-1}$. T_c and $\psi_r(\lambda)$ are defined in Eqs. (2.11b) and (2.14c), respectively.

Proof:

From Eqs. (2.6) in Chapter II and (3.2), we obtain

$$A^{-1}(\lambda) = C(\lambda I_n - A)^{-1}B+D = N_r(\lambda)D_r^{-1}(\lambda)+D = \hat{N}_r(\lambda)D_r^{-1}(\lambda)$$

where $N_r(\lambda) = CT_c^{-1}\psi_r(\lambda)\in C^{m\times m}[\lambda]$; $\hat{N}_r(\lambda) = N_r(\lambda)+DD_r(\lambda)$; T_c is a transformation matrix which transforms (A,B) into the canonical controller pair. Since $A(\lambda)$ is nonsingular; $\hat{N}_r(\lambda)$ must be nonsingular or $A(\lambda) = D_r(\lambda)\hat{N}_r^{-1}(\lambda)$. Also, since $\sigma(A(\lambda)) = \sigma(A) = \sigma(D_r(\lambda))$, we obtain $\det(A(\lambda)) = K_c\det(D_r(\lambda))$, where K_c is a nonzero constant. Therefore, $\det(\hat{N}_r^{-1}(\lambda)) = K_c$. As a result, $\hat{N}_r^{-1}(\lambda) = U_r(\lambda)$ is a unimodular λ-matrix. ∎

Theorem 3.1 provides a method for determining a unimodular λ-matrix which converts a nonsingular column-reduced λ-matrix to the column-reduced canonical λ-matrix.

The dual results for row-reduced λ-matrices are as follows:

Lemma 3.3   Let $A(\lambda)\in C^{p\times p}[\lambda]$ be a row-reduced λ-matrix with row degree $\hat{\nu}_i$, $i=1,\ldots,p$, then $A(\lambda)$ can be represented as

$$A(\lambda) = [D_h(\lambda)-\psi_0(\lambda)A_{0\ell}]D_{0h} \tag{3.5}$$

where D_{0h} is the leading row matrix of $A(\lambda)$ and $D_h(\lambda) \triangleq \text{diag}[\lambda^{\hat{\nu}}, i=1,\ldots,m]$; $\psi_0(\lambda) \triangleq [\psi_{01}(\lambda),\psi_{02}(\lambda),\ldots,\psi_{0p}(\lambda)]^T$; $\psi_{0i}(\lambda) \triangleq [0_{1\times\hat{\tau}_{i-1}},1,\lambda,\ldots,\lambda^{\hat{\nu}_i-1},0_{1\times(n-\tau_i)}]$ for $\hat{\nu}_i>0$ and $\psi_{0i}(\lambda) = 0_{1\times n}$ for $\hat{\nu}_i=0$; $\hat{\tau}_i \triangleq \sum_{j=1}^{i}\hat{\nu}_j, i=1,\ldots,p$; $n \triangleq \hat{\tau}_p$; $A_{0\ell}\in C^{n\times p}$ can be determined from D_{0h} and the coefficients of each entry of $A(\lambda)$. ∎

<u>Lemma 3.4</u> $A^{-1}(\lambda) = \hat{C}_0(\lambda I_n - \hat{A}_0)^{-1}\hat{B}_0 + \hat{D}_0$ (3.6a)

where

(3.6b)

$(A_{0\ell})_i \overset{\Delta}{=}$ The ith column of $A_{0\ell}$, $i=1,\ldots,p$.

$$\hat{B}_0 \overset{\Delta}{=} \psi_0^T(0) \tag{3.6c}$$

$$\hat{C}_0 \overset{\Delta}{=} D_{0h}^{-1}\hat{E}_{c0}; \quad \hat{E}_{c0} \overset{\Delta}{=} [\hat{e}_{c01}, \hat{e}_{c02}, \ldots, \hat{e}_{c0p}]^T; \quad e_{c0i} = \begin{cases} e_n^{\hat{\tau}_i}, & \text{if } \hat{\nu}_i > 0 \\ 0_{n \times 1} & \text{if } \hat{\nu}_i = 0 \end{cases} \tag{3.6d}$$

$$\hat{D}_0 \overset{\Delta}{=} D_{0h}^{-1}[I_p - \psi_0(0)\psi_0^T(0)] \tag{3.6e}$$

$(\hat{A}_0, \hat{B}_0, \hat{C}_0, \hat{D}_0)$ is referred to as an <u>observer</u> (not canonical observer) <u>form</u> <u>realization</u> of $A^{-1}(\lambda)$. In general, if we define $A \overset{\Delta}{=} \hat{T}_0\hat{A}_0\hat{T}_0^{-1}$, $B \overset{\Delta}{=} \hat{T}_0\hat{B}_0$, $C \overset{\Delta}{=} \hat{C}_0\hat{T}_0^{-1}$ and $D = \hat{D}_0$, where $\hat{T}_0$ is a coordinates transformation matrix, then the inverse of a row-reduced λ-matrix, $A(\lambda)$, can be realized by a quadruple (A,B,C,D), i.e.

$$A^{-1}(\lambda) = C(\lambda I_n - A)^{-1}B + D \tag{3.7}$$

Theorem 3.2 Let $A(\lambda)$ be a row-reduced λ-matrix, which can be realized, using a quadruple (A,B,C,D). Let $D_\ell(\lambda)$ be the left characteristic λ-matrix of (A,C). Then

$$A(\lambda) = U_\ell(\lambda)D_\ell(\lambda) \tag{3.8}$$

where $U_\ell(\lambda)$ is a unimodular λ-matrix, $U_\ell(\lambda) = [\psi_\ell(\lambda)T_0^{-1}B + D_\ell(\lambda)D]^{-1}$, T_0 and $\psi_\ell(\lambda)$ are defined in Eqs. (2.17b) and (2.20c), respectively. ∎

Combining Lemma 1.2 and Theorems 3.1 and 3.2 yields the results as follows.

Theorem 3.3 A nonsingular λ-matrix is row or column equivalent to a row-reduced or column-reduced _canonical_ λ-matrix, respectively. ∎

Theorem 3.3 reveals the fact that the structural properties of nonsingular λ-matrices can be studied from the appropriate row-reduced or column-reduced canonical λ-matrices. In the next chapter, the structure theorems of divisors of nonsingular matrices are derived, using the row-reduced or column-reduced canonical λ-matrices.

The following extension of Theorem 3.3 to the RMFD and LMFD can be easily proved.

Theorem 3.4 An irreducible RMFD (LMFD) described by $G(\lambda) = N(\lambda)D^{-1}(\lambda)$, $(G(\lambda) = D^{-1}(\lambda)N(\lambda))$, can be converted to a canonical RMFD (LMFD) described by $G(\lambda) = \hat{N}_r(\lambda)D_r^{-1}(\lambda)$, $(G(\lambda) = D_\ell^{-1}(\lambda)\hat{N}_\ell(\lambda))$, where $D(\lambda)$ and $N(\lambda)$ are coprime and $D_r(\lambda)(D_\ell(\lambda))$ is a column-(row-) reduced canonical λ-matrix. ∎

Corollary 3.1 An irreducible proper RMFD or LMFD can be realized by a canonical controller-form state-space quadruple (A_c, B_c, C_c, D_c) in Eq. (2.12) or a canonical observer-form state-space quadruple (A_0,B_0,C_0,D_0) in Eq. (2.18), respectively.

Proof:

Let $G(\lambda) = N(\lambda)D^{-1}(\lambda)$ be an irreducible, proper RMFD, then from Theorem 3.4 we obtain $G(\lambda) = \hat{N}_r(\lambda)D_r^{-1}(\lambda)$, where $\partial_{ci}(\hat{N}_r(\lambda)) \leq \partial_{ci}(D_r(\lambda))$. Thus, $G(\lambda)$ can be expressed as $G(\lambda) = N_r(\lambda)D_r^{-1}(\lambda)+D_c(\lambda)$, where $\partial_{ci}(N_r(\lambda))<\partial_{ci}(D_r(\lambda))$. Since $G(\lambda)$ is proper, we have $\lim_{\lambda\to\infty} G(\lambda)<\infty$. Also $N_r(\lambda)D_r^{-1}(\lambda)$ is strictly proper, or $\lim_{\lambda\to\infty} N_r(\lambda)D_r^{-1}(\lambda) = 0$, and therefore $\lim_{\lambda\to\infty} D_c(\lambda)<\infty$, or $D_c(\lambda) = D_c$ is a constant matrix. Thus

$$G(\lambda) = N_r(\lambda)D_r^{-1}(\lambda)+D_c .$$

From Eqs. (2.14) and (2.16), we can find the minimal realization quadruple (A_c,B_c,C_c,D_c) in the canonical controller form.

The minimal realization of the irreducible proper LMFD in the canonical observer form can be proved in a similar way. ∎

3.2 Latent Structure of Nonsingular λ-matrices

As stated in Theorem 3.3 in Section 3.1, a nonsingular λ-matrix is row or column equivalent to a row-reduced or column-reduced canonical λ-matrix, respectively. We shall first concentrate on the latent structures of canonical λ-matrices and then extend the results to nonsingular λ-matrices.

The latent roots of a nonsingular λ-matrix $D(\lambda)$ are defined as the roots of $\det(D(\lambda)) = 0$.

Lemma 3.5 [17-19] Let the set of latent roots of $D(\lambda)$ be $\sigma(D(\lambda))$, and the set of eigenvalues of A, which is the system map of a minimal realization of $D^{-1}(\lambda)$, be $\sigma(A)$. Then $\sigma(D(\lambda)) = \sigma(A)$, and

$$|\sigma(D(\lambda))| = |\sigma(A)| = n .$$

where $|\;\;|$ denotes the cardinality of the set.

Proof:

Directly from Theorem 3.3 we have

$$D(\lambda) = D_r(\lambda)U_r(\lambda) \qquad (3.10a)$$

where $D_r(\lambda)$ is a column-reduced canonical λ-matrix, and $U_r(\lambda)$ is unimodular. From Lemma 2.1, we have

$$D^{-1}(\lambda) = U_r^{-1}(\lambda)[C_r(\lambda I_n - A)^{-1}B + D_r] \qquad (3.10b)$$

Thus the results of Lemma 3.5 follow. ∎

The generalized latent vectors of a nonsingular λ-matrix are defined as follows:

Definition 3.2 [17-19] Let λ_i be a latent root of $D(\lambda)$. A left Jordan chain of left generalized latent vectors (left Jordan chain for abbreviation) for $D(\lambda)$ associated with λ_i is a set of nonzero vectors p_{ij}, $0 \le j \le \ell_i - 1$ such that

$$\sum_{k=0}^{j} \frac{1}{k!} (D^{(k)}(\lambda_i))^T p_{i(j-k)} = 0_{m \times 1} \qquad (3.11a)$$

where $D^{(k)}(\lambda) \triangleq d^{(k)}D(\lambda)/d\lambda^k = $ k-th derivative of $D(\lambda)$ with respect to λ (3.11b)

The scalar ℓ_i is named the length of the Jordan chain [41]. Similarly, a right Jordan chain of $D(\lambda)$ associated with λ_i is a set of nonzero vectors q_{ij},

$0 \leq j \leq \ell_i - 1$ such that

$$\sum_{k=0}^{j} \frac{1}{k!} (D^{(k)}(\lambda_i)) q_{i(j-k)} = 0_{m \times 1} \qquad (3.11c)$$

The vectors p_{i0} and q_{i0} are referred to as the <u>primary</u> left and right latent vectors, respectively. $\qquad \qquad \Box$

We shall first present the properties of left/right Jordan chains of the column-reduced canonical λ-matrices, and the relations to the eigenvectors of the system maps in their associated minimal realization quadruples.

The salient properties of left and right Jordan chains of a column-reduced canonical λ-matrix $D_r(\lambda)$ are described as follows.

<u>Lemma 3.6</u> Let $(q_{ij})_\ell$ be the ℓ-th component of q_{ij}, then for $0 \leq \ell \leq m$ we have $(q_{ij})_\ell = 0$ if the Kronecker index, of a minimal realization quadruple of $D_r^{-1}(\lambda)$, satisfies $\kappa_\ell = 0$, $0 \leq \ell \leq m$.

<u>Proof:</u>

From Eq. (3.11c) we have

$$D_r(\lambda_i) q_{i0} = D_{rh} D(\lambda_i) q_{i0} = 0_{m \times 1} \ .$$

Since D_{rh} is nonsingular, and if $\kappa_\ell = 0$, the ℓth column of $D_r(\lambda_i)$ is e_ℓ^m and ℓth row of $D_r(\lambda_i)$ is $(e_\ell^m)^T$. Thus, the ℓth component of q_{i0} is zero. Also, from

$$D_r(\lambda_i) q_{i1} + D_r^{(1)}(\lambda_i) q_{i0} = 0_{m \times 1}$$

We find that the ℓth component of q_{i1} is zero. Repeatedly using Eq. (3.11c) results in $(q_{ij})_\ell = 0$ if $\kappa_\ell = 0$, $1 \leq \ell \leq m$. $\qquad \blacksquare$

Lemma 3.7 If P_{ij}, $0 \leq j \leq \ell_i - 1$, is a left Jordan chain of $D_r(\lambda)$ and $\hat{P}_{ij} \triangleq D_{rh}^T P_{ij}$, then the ℓ-th component, $(\hat{P}_{ij})_\ell$, of $\hat{P}_{ij}$ is zero. This implies that the ℓ-th column of D_{rh} is orthogonal to P_{ij} if $\kappa_\ell = 0$ for $0 \leq \ell \leq m$. ∎

Lemma 3.7 can be easily proved by observing the fact that $\hat{P}_{ij}$, $0 \leq j \leq \ell_i - 1$, is a right Jordan chain of $D_r^T(\lambda)$ and the ℓ-th component of $\hat{P}_{ij}$ is zero if $\kappa_\ell = 0$.

Let (A_c, B_c, C_c, D_c) be a minimal realization quadruple of the inverse of a column-reduced canonical λ-matrix $D_r(\lambda)$, and let (A_c, B_c) be a controller canonical pair as defined in Eqs. (2.13a) and (2.13d), i.e.

$$D_r^{-1}(\lambda) = C_c(\lambda I_n - A_c)^{-1} B_c + D_c \tag{3.12}$$

The relationships between the Jordan chains of A_c and Jordan chains of $D_r(\lambda)$ can be described in the following important theorems.

Theorem 3.5 If q_{ij}, $0 \leq j \leq \ell_i - 1$, is a right Jordan chain of $D_r(\lambda)$ associated with a latent root λ_i, then q_{cij}, $0 \leq j \leq \ell_i - 1$, a right Jordan chain of A_c associated with an eigenvalue λ_i, can be found as

$$q_{cij} = \sum_{k=0}^{j} \frac{1}{k!} \psi_r^{(k)}(\lambda_i) q_{i(j-k)} \tag{3.13a}$$

where

$$\psi_r^{(k)}(\lambda) \triangleq d^{(k)} \psi_r(\lambda)/d\lambda^k = \text{n-th derivate of } \psi_r(\lambda) \text{ with respect to } \lambda \tag{3.13b}$$

and $\psi_r(\lambda)$ is defined in Eq. (2.14c).

Proof:

From Eqs. (2.14b) and (2.15b) we have

$$B_c D_r(\lambda) = (\lambda I_n - A_c)\psi_r(\lambda) \tag{3.13c}$$

Differentiating both sides of Eq. (3.13c) k times yields

$$B_c D_r^{(k)}(\lambda) = k\psi_r^{(k-1)}(\lambda) + (\lambda I_n - A_c)\psi_r^{(k)}(\lambda) , \quad k \leq 1 \tag{3.13d}$$

If q_{ij}, $0 \leq j \leq \ell_i - 1$, is a right Jordan chain of $D_r(\lambda)$, from Eqs. (3.11c) and (3.13d) we have

$$B_c \sum_{k=0}^{j} \frac{1}{k!} D_r^{(k)}(\lambda_i) q_{i(j-k)} = \sum_{k=1}^{j} \frac{1}{(k-1)!} [\psi_r^{(k-1)}(\lambda_i) q_{i(j-k)}]$$

$$+ \sum_{k=0}^{j} \frac{1}{k!} [(\lambda_i I_n - A_c)\psi_r^{(k)}(\lambda_i) q_{i(j-k)}]$$

$$= 0_{n \times 1}, \quad 1 \leq j \leq \ell_i - 1 \tag{3.14a}$$

Thus, we have

$$(\lambda_i I_n - A_c)\psi_r(\lambda_i) q_{i0} = 0_{n \times 1} \tag{3.14b}$$

$$(\lambda_i I_n - A_c) \sum_{k=0}^{j} \frac{1}{k!} \psi_r^{(k)}(\lambda_i) q_{i(j-k)} + \sum_{k=0}^{j-1} \frac{1}{k!} \psi_r^{(k)}(\lambda_i) q_{i(j-1-k)} = 0_{n \times 1}, \quad 1 \leq j \leq \ell_i - 1 \tag{3.14c}$$

Define $q_{cij} \triangleq \sum_{k=0}^{j} \frac{1}{k!} \psi_r^{(k)}(\lambda_i) q_{i(j-k)}$ $\tag{3.15a}$

Then, Eq. (3.14) becomes

$$(\lambda_i I_n - A_c) q_{ci0} = 0_{n \times 1} \tag{3.15b}$$

$$(\lambda_i I_n - A_c)q_{cij} + q_{ci(j-1)} = 0_{n\times 1} \ , \quad 1 \leq j \leq \ell_i - 1 \tag{3.15c}$$

Obviously q_{cij}, $0 \leq j \leq \ell_i - 1$, is a Jordan chain of A_c. ∎

__Theorem 3.6__ If q_{cij}, $0 \leq j \leq \ell_i - 1$, is a right Jordan chain of A_c associated with an eigenvalue λ_i, then q_{ij}, a right Jordan chain of $D_r(\lambda)$, can be determined by

$$q_{ij} = \psi_r^T(0)q_{cij} \ , \quad j = 0,1,\ldots,\ell_i - 1 \tag{3.16}$$

__Proof:__

From Eq. (3.15a) we have $q_{ci0} = \psi_r(\lambda_i)q_{i0}$. So from Eq. (2.23b),
$$\psi_r^T(0)q_{ci0} = \psi_r^T(0)\psi_r(\lambda_i)q_{i0} = \psi_r^T(0)\psi_r(0)q_{i0}$$

The ℓ-th component of $\psi_r^T(0)q_{ci0}$ can be expressed as

$$(\psi_r^T(0)q_{ci0})_\ell = \begin{cases} (q_{i0})_\ell \ , & \kappa_\ell > 0 \\ 0 \ , & \kappa_\ell = 0 \end{cases}$$

From Lemma 3.6 we have $q_{i0} = \psi_r^T(0)q_{ci0}$. Thus, from Eq. (3.15a) we have

$$\psi_r^T(0)q_{cij} = \sum_{k=0}^{j} \frac{1}{k!} \psi_r^T(0)\psi_r^{(k)}(\lambda_i)q_{i(j-k)} \ , \quad j > 0$$

Since

$$\psi_r^T(0)\psi_r^{(k)}(\lambda_i) = \begin{cases} \psi_r^T(0)\psi_r(0) & \text{if } k=0 \\ 0_m & \text{if } k>0 \end{cases}$$

This implies

$$\psi_r^T(0)q_{cij} = \psi_r^T(0)\psi_r(0)q_{ij}$$

The ℓ-th component of $\psi_r^T(0)q_{cij}$ is

$$(\psi_r^T(0)q_{cij})_\ell = \begin{cases} (q_{ij})_\ell & \text{if } \kappa_\ell > 0 \\ 0 & \text{if } \kappa_\ell = 0 \end{cases}$$

Thus, we have the result in Eq. (3.16). ∎

<u>Theorem 3.7</u> If p_{cij}, $0 \leq j \leq \ell_i - 1$, is a left Jordan chain of A_c associated with an eigenvalue λ_i, then p_{ij}, a left Jordan chain of $D_r(\lambda)$, can be determined by

$$p_{ij} = B_c^T p_{cij} , \quad 0 \leq j \leq \ell_i - 1 \tag{3.17}$$

<u>Proof:</u>

From Eq. (3.13d), we have

$$[D_r^{(k)}(\lambda)]^T B_c^T = k[\psi_r^{(k-1)}(\lambda)]^T + [\psi_r^{(k)}(\lambda)]^T (\lambda I_n - A_c)^T .$$

Since p_{cij}, $0 \leq j \leq \ell_i - 1$ is a left Jordan chain of A_c, we have

$$(\lambda_i I_n - A_c)^T p_{cij} = -p_{ci(j-1)}; \quad 0 \leq i \leq \ell_i - 1 ,$$

and $p_{ci(-1)} \triangleq 0_{n \times 1}$.

Therefore,

$$\sum_{k=0}^{j} \frac{1}{k!} [D_r^{(k)}(\lambda)]^T B_c^T p_{ci(j-k)} = \sum_{k=1}^{j} \frac{1}{(k-1)!} [\psi_r^{(k-1)}(\lambda_i)]^T p_{ci(j-k)}$$

$$+ \sum_{k=0}^{j} \frac{1}{k!} [\psi_r^{(k)}(\lambda_i)]^T (\lambda_i I_n - A_c)^T P_{ci(j-k)}$$

$$= \sum_{k=0}^{j-1} \frac{1}{k!} [\psi_r^{(k)}(\lambda_i)]^T P_{ci(j-k-1)} - \sum_{k=0}^{j} \frac{1}{k!} [\psi_r^{(k)}(\lambda_i)]^T P_{ci(j-k-1)}$$

$$= - \frac{1}{j!} [\psi_r^{(j)}(\lambda_i)]^T P_{ci(-1)} = 0_{m \times 1}$$

Thus, we conclude that $B_c^T P_{cij}$, $0 \le j \le \ell_i - 1$, is a left Jordan chain of $D_r(\lambda)$. ∎

Theorem 3.8   If P_{ij}, $0 \le j \le \ell_i - 1$, is a left Jordan chain of $D_r(\lambda)$ associated with latent root λ_i, then a left Jordan chain P_{cij} of A_c can be determined by

$$P_{cij} = T_p \sum_{k=0}^{j} \frac{1}{k!} \psi_r^{(k)}(\lambda_i) D_{rh}^T P_{i(j-k)} \tag{3.18}$$

where $T_p = \{[T_p]_{ij}\}$; $1 \le j \le m$, $1 \le i \le m$ and $\kappa_i \kappa_j \ne 0$;

$[T_p]_{ii}$ = reversed upper triangular Toeplitz matrix [11] with first column $[\bar{a}_{ii2}, \ldots, \bar{a}_{ii\kappa_i}, 1]^T$;

$[T_p]_{ji}$ = reversed upper triangular Toeplitz matrix with first column $[\bar{a}_{ij2}, \ldots, \bar{a}_{ij\kappa_i}, 0]^T$ if $\kappa_i \le \kappa_j$, and $[\bar{a}_{ij2}, \ldots, \bar{a}_{ij\kappa_j}, 0, \ldots, 0]^T$ if $\kappa_i > \kappa_j$.

Proof:

From Lemma 3.7, $D_{rh}^T P_{ij}$, $0 \le j \le \ell_i - 1$, are generalized left latent vectors of a left Jordan chain of $\sigma(\lambda) = D_{rh}^{-1} D_r(\lambda)$. Define

$$\hat{P}_{cij} \triangleq \sum_{k=0}^{j} \frac{1}{k!} \psi_r^{(k)}(\lambda_i) D_{rh}^T P_{i(j-k)}, \quad 0 \le j \le \ell_i - 1$$

It follows from Theorem 3.5 that $\hat{p}_{cij}$, $0 \leq j \leq \ell_i - 1$ is a right Jordan chain of $\hat{A}_c$ which is the associated companion form for $D^T(\lambda)$, and $\hat{A}_c \triangleq (\hat{A}_c)_{ij}$ for $1 \leq i \leq m$ and $1 \leq j \leq m$ is defined according to:

$$(\hat{A}_c)_{ii} = (A_c)_{ii} \in R^{\kappa_i \times \kappa_i}$$

$$(\hat{A}_c)_{ji} = \begin{bmatrix} 0_{(\kappa_j - 1) \times k_i} \\ \hat{A}_{ji} \end{bmatrix} \in R^{\kappa_j \times \kappa_i}; \hat{A}_{ji} = \begin{cases} [\bar{a}_{ij1}, \ldots, -\bar{a}_{ij\kappa_i}] & \text{if } \kappa_i \leq \kappa_j \\ [-\bar{a}_{ij1}, \ldots, -\bar{a}_{ij\kappa_j}, 0, \ldots, 0] & \text{if } \kappa_i > \kappa_j; i \neq j \end{cases}$$

By direct computations, it can be easily shown that $A_c^T = T_p \hat{A}_c T_p^{-1}$. Thus, a left Jordan chain of A_c^T is $p_{cij} = T_p \hat{p}_{cij}$. ■

From Theorems 3.5–3.8, the relationships between the Jordan chains of $D_r(\lambda)$ and the Jordan chains of A, which is a system map of an arbitrary minimal realization quadruple of $D_r^{-1}(\lambda)$, can be formulated as follows:

<u>Corollary 3.2</u> Let $D_r(\lambda)$ be the right characteristic λ-matrix of a reachable pair (A,B); q_{aij} for $0 \leq j \leq \ell_i - 1$ be a right Jordan chain of A associated with an eigenvalue λ_i of A; q_{ij} for $0 \leq j \leq \ell_i - 1$ be a right Jordan chain of $D_r(\lambda)$ associated with the same latent root λ_i of $D_r(\lambda)$. Then q_{aij} and q_{ij} can be related by

$$q_{aij} = T_c^{-1} \sum_{k=0}^{j} \frac{1}{k!} \psi_r^{(k)}(\lambda_i) q_{i(j-k)}, \quad 0 \leq j \leq \ell_i - 1$$

$$q_{ij} = \psi_r^T(0) T_c q_{aij}, \quad 0 \leq j \leq \ell_i - 1$$

where $\psi_r^{(k)}(\lambda)$ is defined in Eq. (3.13b), and T_c is the transformation matrix which transforms A to its controller canonical form $A_c = T_c A T_c^{-1}$. ■

Proof:

Corollary 3.2 can be proved in a straightforward fashion from Theorems 3.5 and 3.6. In a similar way using Theorems 3.7 and 3.8 we can obtain:

Corollary 3.3   Let $D_r(\lambda)$ be the characteristic λ-matrix of a reachable pair (A,B); P_{aij} for $0 \leq j \leq \ell_i - 1$ be a left Jordan chain of A associated with the eigenvalue λ_i of A; p_{ij} for $0 \leq j \leq \ell_i - 1$ be a left Jordan chain of $D_r(\lambda)$ associated with the same latent root λ_i of $D_r(\lambda)$. Then, P_{aij} and p_{ij} can be related by

$$P_{aij} = T_c^T T_p \sum_{k=0}^{j} \frac{1}{k!} \psi_r^{(k)}(\lambda_i) D_{rh}^T P_{i(j-k)}, \quad 0 \leq j \leq \ell_i - 1$$

and

$$P_{ij} = B^T P_{cij}, \quad 0 \leq j \leq \ell_i - 1$$

where $\psi_r^{(k)}(\lambda)$ is defined in Eq. (3.13b) and T_c is the transformation matrix such that $A_c = T_c A T_c^{-1}$, which is the controller canonical form of A.   ∎

It is well known that the Jordan chains of a matrix are not unique; the same is true for the Jordan chains of a λ-matrix. However, from Theorems 3.5 through 3.8, we observe that, given a left/right Jordan chain of a system map A_c, the corresponding left/right Jordan chain of the right characteristic λ-matrix of a reachable pair (A_c, B_c) can be uniquely determined, and vice versa. Also, from Theorems 3.5 through 3.8, we observe that $\psi_r(\lambda)$ in Eq. (2.14c) links the generalized eigenvectors of A_c to the generalized latent vectors of $D_r(\lambda)$. Moreover, the input matrix B_c in Eq. (2.13d) or B in Eq. (2.1) relates the left generalized eigenvectors of A_c to the left generalized latent vector of $D_r(\lambda)$.

Let M_c be a modal matrix of A_c. Then, we have $A_J \triangleq M_c^{-1} A_c M_c$, where A_J is a system map in Jordan block canonical form, and M_c and (M_c^{-1}) contain all the right and left generalized eigenvectors of A_c. Thus, from Theorems 3.6 and 3.7 we can easily derive the following results.

<u>Lemma 3.8</u> $(M_c^{-1}B_c)^T$ contains all the left Jordan chains of a column-reduced canonical λ-matrix $D_r(\lambda)$, and $\psi_r^T(0)M_c$ contains all the right Jordan chains of $D_r(\lambda)$. ∎

<u>Corollary 3.4</u>   Let (A,B) be a reachable pair, M be a modal matrix of A, and $D_r(\lambda)$ be the characteristic λ-matrix of (A,B). Then, $(M^{-1}B)^T$ and $\psi_r^T(0)M$ contain all the left and right Jordan chains of $D_r(\lambda)$, respectively. ∎

Now, from Lemmas 2.1 and 3.8 we have the following conclusion.

<u>Theorem 3.9</u>   Let $(A_J,B_J,\overline{C}_J,\overline{D}_J)$ be a minimal realization of a column-reduced canonical λ-matrix $D_r(\lambda)$ with A_J in Jordan form. Then,

$$D_r^{-1}(\lambda) = \overline{C}_J(\lambda I_n - A_J)^{-1}B_J + \overline{D}_J\ , \tag{3.19}$$

where B_J^T in Eq. (3.19) contains all the left Jordan chains of $D_r(\lambda)$, and $\overline{C}_J$ contains all the right Jordan chains of $D_r(\lambda)$.

<u>Proof:</u>

From Lemma 2.1, $(A,B,\overline{C},\overline{D})$ is a minimal realization of $D_r(\lambda)$. Thus

$$D_r^{-1}(\lambda) = \overline{C}(\lambda I_n - A)^{-1}B + \overline{D} = \psi_r^T(0)(\lambda I_n - A_c)^{-1}B_c + \overline{D}$$

$$= \psi_r^T(0)M_c(\lambda I_n - A_J)^{-1}M_c^{-1}B_c + \overline{D} \triangleq \overline{C}_J(\lambda I_n - A_J)^{-1}B_J + \overline{D}_J$$

where $\overline{C}_J = \psi_r^T(0)M_c$; $B_J = M_c^{-1}B_c$; $\overline{D}_J = \overline{D}$.

Hence, from Lemma 3.8, B_J^T and $\overline{C}_J$ contain all the left and right Jordan chains of $D_r(\lambda)$. ∎

In an analogous fashion, we can easily obtain the following results for a latent structure of the row-reduced canonical λ-matrices:

__Theorem 3.10__   Let $(A_J, \bar{B}_J, C_J, \bar{D}_J)$ be a minimal realization of a row-reduced canonical λ-matrix, $D_\ell(\lambda)$, with A_J in a Jordan form.   Then,

$$D_\ell^{-1}(\lambda) = C_J(\lambda I_n - A_J)^{-1}\bar{B}_J + \bar{D}_J \tag{3.20}$$

where $\bar{B}_J^T$ in Eq. (3.20) contains all the left Jordan chains of $D_\ell(\lambda)$, and C_J contains all the right Jordan chains of $D_\ell(\lambda)$. ■

The latent structures of general nonsingular λ-matrices can be derived directly from Theorems 3.3, 3.9, and 3.10:

__Theorem 3.11__   Let $D(\lambda)$ be a nonsingular λ-matrix and

$$D(\lambda) = D_r(\lambda)U_r(\lambda)$$

where $D_r(\lambda)$ is a column-reduced canonical λ-matrix and $U_r(\lambda)$ is unimodular.   Let λ_i be a latent root of $D_r(\lambda)$, p_{ij} and q_{ij} for $0 \le j \le \ell_i - 1$ be an associated left and an associated right Jordan chain of $D_r(\lambda)$ with length ℓ_i, respectively.   Then

$$\bar{p}_{ij} \overset{\Delta}{=} p_{ij}, \qquad 0 \le j \le \ell_i - 1 \tag{3.21a}$$

and

$$\bar{q}_{ij} \overset{\Delta}{=} \sum_{k=0}^{j} \frac{1}{k!} U_r^{-(k)}(\lambda_i) q_{i(j-k)}, \qquad 0 \le j \le \ell_i - 1 \tag{3.21b}$$

are respectively a left and a right Jordan chain of $D(\lambda)$ with length ℓ_i.   These correspond to the latent root λ_i of $D(\lambda)$. $U_r^{-(k)}(\lambda_i)$ denotes the value of the k-th derivative of $U_r^{-1}(\lambda)$ evaluated at $\lambda = \lambda_i$.

58

Proof:

From Theorem 3.3, we have

$$D(\lambda) = D_r(\lambda)U_r(\lambda)$$

Since p_{ij}, $0 \leq j \leq \ell_i - 1$, is a left Jordan chain of $D_r(\lambda)$, from Definition 3.2 we have

$$\sum_{k=0}^{j} \frac{1}{k!} (D_r^{(k)}(\lambda_i))^T p_{i(j-k)} = 0_{m \times 1}, \qquad 0 \leq j \leq \ell_i - 1 \tag{3.22}$$

Since $D(\lambda) = D_r(\lambda)U_r(\lambda)$, we have

$$D_r^{(k)}(\lambda) = [D(\lambda)U_r^{-1}(\lambda)]^k$$

$$= \sum_{s=0}^{k} {}_kC_s D^{(s)}(\lambda)U_r^{-(k-s)}(\lambda) \tag{3.23a}$$

where ${}_kC_s$ is a binomial coefficient,

$${}_kC_s \triangleq \frac{k!}{s!(k-s)!} \tag{3.23b}$$

Substituting Eq. (3.23) into (3.22) yields

$$\sum_{k=0}^{j} \frac{1}{k!} (\sum_{s=0}^{k} {}_kC_s [D^{(s)}(\lambda_i)U_r^{-(k-s)}(\lambda_i)U_r^{-(k-s)}(\lambda_i)]^T p_{i(j-k)} = 0_{m \times 1} \qquad 0 \leq j \leq \ell_i - 1 \tag{3.24a}$$

Equation (3.24a) can be rearranged as

$$\sum_{s=0}^{j} \frac{1}{s!} [U_r^{-(s)}(\lambda_i)]^T \sum_{k=s}^{j} \frac{1}{(k-s)!} [D^{(k-s)}(\lambda_i)]^T p_{i(j-k)} = 0_{m \times 1} \qquad 0 \leq j \leq \ell_i - 1 \tag{3.24b}$$

or

$$\sum_{s=0}^{j} \frac{1}{s!} \ [U_r^{-(s)}(\lambda_i)]^T \sum_{k=0}^{j-s} \frac{1}{k!} \ [D^{(k)}(\lambda_i)]^T P_{i(j-s-k)} = 0_{m \times 1} \qquad 0 \le j \le \ell_i - 1$$

(3.24c)

Since $U_r^{-1}(\lambda_i)$ is nonsingular for all λ_i, from Eq. (3.24c) we conclude that

$$\sum_{k=0}^{j} \frac{1}{k!} \ [D^{(k)}(\lambda_i)]^T P_{i(j-k)} = 0_{m \times 1} \qquad 0 \le j \le \ell_i - 1$$

(3.25)

By Definition 3.2, P_{ij} for $0 \le j \le \ell_i - 1$ is a left Jordan chain of $D(\lambda)$ corresponding to the latent root λ_i.

Furthermore, since q_{ij} for $0 \le j \le \ell_i - 1$ is a right Jordan chain of $D_r(\lambda)$, from Definition 3.2 we have

$$\sum_{k=0}^{j} \frac{1}{k!} \ [D_r^{(k)}(\lambda_i)] q_{i(j-k)} = 0_{m \times 1} \qquad 0 \le j \le \ell_i - 1$$

(3.26)

Substituting Eq. (3.23) into (3.26) yields

$$\sum_{k=0}^{j} \frac{1}{k!} \ [\sum_{s=0}^{k} {}_kC_s D^{(s)}(\lambda_i) U_r^{-(k-s)}(\lambda_i)] q_{i(j-k)} = 0_{m \times 1} \qquad 0 \le j \le \ell_i - 1$$

(3.27a)

Equation (3.27a) can be rearranged as

$$\sum_{s=0}^{j} \frac{1}{s!} D^{(s)}(\lambda_i) \sum_{k=0}^{j-s} \frac{1}{k!} U_r^{-(k)}(\lambda_i) q_{i(j-s-k)} = 0_{m \times 1} \qquad 0 \le j \le \ell_i - 1$$

(3.27b)

By Definition 3.2, we conclude that

$$\bar{q}_{ij} \triangleq \sum_{k=0}^{j} \frac{1}{k!} U_r^{-(k)}(\lambda_i) q_{i(j-k)} \quad \text{for } 0 \leq j \leq \ell_i - 1$$

is a right Jordan chain of $D(\lambda)$ corresponding to the latent root λ_i. ∎

From Theorem 3.11, we can easily deduce the following:

<u>Corollary 3.5</u> Let $D(\lambda)$ and $D_r(\lambda)$ be defined in Theorem 3.11, $\bar{p}_{ij}$ and $\bar{q}_{ij}$ for $0 \leq j \leq \ell_i - 1$ be a left and a right Jordan chain of $D(\lambda)$ corresponding to a latent root λ_i. Then

$$p_{ij} \triangleq \bar{p}_{ij} , \qquad 0 \leq j \leq \ell_i - 1 \tag{3.28a}$$

and

$$q_{ij} \triangleq \sum_{k=0}^{j} \frac{1}{k!} U_r^{(k)}(\lambda_i) \bar{q}_{i(j-k)} , \qquad 0 \leq j \leq \ell_i - 1 \tag{3.28b}$$

are a left and a right Jordan chain of $D_r(\lambda)$ corresponding to the same latent root λ_i. ∎

Again, for completeness we state the result corresponding to Theorem 3.11 and Corollary 3.5 for the row-reduced case.

<u>Theorem 3.12</u> Let $D(\lambda)$ be a nonsingular λ-matrix, and

$$D(\lambda) = U_\ell(\lambda) D_\ell(\lambda)$$

where $D_\ell(\lambda)$ is a row-reduced canonical λ-matrix, and $U_\ell(\lambda)$ is unimodular. Let p_{ij} and q_{ij} for $0 \leq j \leq \ell_i - 1$ be a left and a right Jordan chain of $D_\ell(\lambda)$ with length ℓ_i, respectively. They correspond to a latent root λ_i of $D_r(\lambda)$. Then

$$\bar{p}_{ij} \triangleq \sum_{k=0}^{j} \frac{1}{k!} [U_\ell^{-(k)}(\lambda_i)]^T p_{i(j-k)} , \qquad 0 \leq j \leq \ell_i - 1 \tag{3.29a}$$

and

$$\bar{q}_{ij} \triangleq q_{ij} , \qquad 0 \leq j \leq \ell_i - 1 \tag{3.29b}$$

are a left and a right Jordan chain of $D(\lambda)$ with length ℓ_i, respectively. These correspond to the latent root λ_i of $D(\lambda)$. ∎

Corollary 3.6   Let $D(\lambda)$ and $D_\ell(\lambda)$ be defined in Theorem 3.12, $\bar{p}_{ij}$ and $\bar{q}_{ij}$ for $0 \leq j \leq \ell_i - 1$ be a left and a right Jordan chain of $D(\lambda)$ corresponding to a latent root λ_i, then

$$p_{ij} \triangleq \sum_{k=0}^{j} \frac{1}{k!} [U_\ell^{(k)}(\lambda_i)]^T \bar{p}_{i(j-k)} , \qquad 0 \leq j \leq \ell_i - 1 \tag{3.30a}$$

and

$$q_{ij} \triangleq \bar{q}_{ij}, \qquad 0 \leq j \leq \ell_i - 1 \tag{3.30b}$$

are a left and a right Jordan chain of $D_\ell(\lambda)$ corresponding to the same latent root λ_i. ∎

Theorems 3.11 and 3.12 with Corollaries 3.2 and 3.3 build up the latent structures of nonsingular λ-matrices based on the latent structures of canonical λ-matrices, whose importance is thereby emphasized.

3.3   Solvents of Nonsingular λ-matrices

In this section we shall extend the idea of latent roots in nonsingular λ-matrices to "matrix roots" which in the literature [17,18] are usually called solvents.

Let $f:C \to C$ be analytic within, and on a simple closed contour C, and x be any point interior to C, then the well-known Cauchy's integral theorem [42] gives

$$f(x) = \frac{1}{2\pi i} \oint_C \frac{f(\lambda)}{\lambda - x} d\lambda \qquad (3.31)$$

where C is traversed in the positive (counter-clockwise) sense.

Using this approach, a class of matrix functions can be generated from scalar function $f(\lambda)$ [43]. Let $X \epsilon C^{m \times m}$, the set of m×m complex matrices, and let $\lambda_1, \ldots, \lambda_q$, for $q \leq m$, be the distinct eigenvalues of X. Suppose that $\lambda_1, \ldots, \lambda_q$ are interior to C, then the matrix function $f(X)$ can be defined [9] as

$$f(X) = \frac{1}{2\pi i} \oint_C f(\lambda)(\lambda I_m - X)^{-1} d\lambda = \frac{1}{2\pi i} \oint_C (\lambda I_m - X)^{-1} f(\lambda) d\lambda \qquad (3.32)$$

where I_m is an m×m identity matrix.

If $f(\lambda)$, an n-th degree complex polynomial with coefficients f_k, is described by

$$f(\lambda) = \sum_{k=0}^{n} f_k \lambda^{n-k} = \sum_{k=0}^{n} \lambda^{n-k} f_k \qquad (3.33a)$$

then, the corresponding matrix function becomes

$$f(X) = \sum_{k=0}^{n} f_k X^{n-k} = \sum_{k=0}^{n} X^{n-k} f_k \qquad (3.33b)$$

where $X^0 \triangleq I_m$.

An n-th degree, m-th order nonsingular λ-matrix $A(\lambda) \epsilon C^{m \times m}[\lambda]$ can be written as follows:

$$A(\lambda) = A_0 \lambda^n + A_1 \lambda^{n-1} + \ldots + A_{n-1} \lambda + A_n \qquad (3.34)$$

where $A_k \epsilon C^{m \times m}$, $k = 0, 1, \ldots, n$, $\det(A(\lambda)) \neq 0$.

Since $A(\lambda)$ is analytic everywhere in C , we can define the right matrix polynomial $A_R : C^{m \times m} \to C^{m \times m}$ as

$$A_R(X) = \frac{1}{2\pi i} \oint_C A(\lambda)(I_m \lambda - X)^{-1} d\lambda \tag{3.35a}$$

where $X \epsilon C^{m \times m}$ has all its eigenvalues interior to the simple closed contour C.

Substituting Eq. (3.34) into Eq. (3.35a) yields

$$A_R(X) = \frac{1}{2\pi i} \oint_C (\sum_{k=0}^{n} A_k \lambda^{n-k})(I_m \lambda - X)^{-1} d\lambda$$

$$= \sum_{k=0}^{n} A_k [\frac{1}{2\pi i} \oint_C \lambda^{n-k}(I_m \lambda - X)^{-1} d\lambda]$$

$$= \sum_{k=0}^{n} A_k X^{n-k} \tag{3.35b}$$

Similarly, the left matrix polynomial of $A(\lambda)$ is defined as

$$A_L(X) = \frac{1}{2\pi i} \oint_C (I_m \lambda - X)^{-1} A(\lambda) d\lambda = \sum_{k=0}^{n} [\frac{1}{2\pi i} \oint_C (I_m \lambda - X)^{-1} \lambda^{n-k} d\lambda] A_k$$

$$= \sum_{k=0}^{n} X^{n-k} A_k \tag{3.36}$$

Having these definitions of right (left) matrix polynomials in Eqs. (3.35) and (3.36), we are ready to define the solvents of λ-matrices.

Definition 3.3 Let R be an m×m complex matrix such that

$$A_R(R) = A_0 R^n + A_1 R^{n-1} + \ldots + A_{n-1} R + A_n = 0_m \tag{3.37}$$

Then R is a <u>right</u> <u>solvent</u> of the λ-matrix A(λ), where 0_m is an m×m null matrix. Similarly, if L is an m×m complex matrix such that

$$A_L(L) = L^n A_0 + L^{n-1} A_1 + \ldots + L A_{n-1} + A_n = 0_m \qquad (3.38)$$

then L is a <u>left</u> <u>solvent</u> of the λ-matrix A(λ). □

The solvents of A(λ) can be constructed by using the latent roots and generalized latent vectors of A(λ). The structure and existence of solvents have been extensively studied and reported in the literature (for example, [17-19, 23-25]). Numerical methods to compute solvents can be found in [24,44-46]. Some important results on the properties of solvents are summarized as follows.

<u>Theorem 3.13</u> If A(λ) has m linearly independent right generalized latent vectors $\hat{q}_1, \ldots, \hat{q}_m$ (left latent vectors $\hat{p}_1, \ldots, \hat{p}_m$) corresponding to latent roots $\lambda_1, \ldots, \lambda_m$, then $Q \Lambda Q^{-1} (P^{-1} \Lambda P)$ is a right (left) solvent, where $P = (\hat{p}_1, \ldots, \hat{p}_m)^T$ $(Q = (\hat{q}_1, \ldots, \hat{q}_m))$ and $\Lambda = \text{diag}(\lambda_1, \ldots, \lambda_m)$. ∎

A set of right (left) solvents R_k, k = 1,...,n (L_k, k = 1,...,n) is called a <u>complete</u> set of right (left) solvents of A(λ) if $\sigma(A(\lambda)) = \bigcup_{k=1}^{n} \sigma(R_k) (\sigma(A(\lambda)) = \bigcup_{k=1}^{n} \sigma(L_k))$, where $\sigma(A(\lambda))$ is the spectrum of A(λ) and $\sigma(R_k)(\sigma(L_k))$ is the spectrum of the right (left) solvent $R_k(L_k)$. The existence of the complete set of right (left) solvents has been investigated by Gohberg et. al. [20,21], and Dennis et. al. [18].

A right solvent R of A(λ) is <u>regular</u> [17] if

$$\sigma(R) \cap \sigma(A^{(1)}(\lambda)) = \phi \qquad (3.39a)$$

where $A^{(1)}(\lambda)$ is given by

$$A(\lambda) = A^{(1)}(\lambda)(I_m\lambda - R) \qquad (3.39b)$$

and ϕ denotes the null set. Similarly, the definition of a regular left solvent is given by

$$\sigma(L) \cap \sigma(\hat{A}^{(1)}(\lambda)) = \phi \qquad (3.40a)$$

where $\hat{A}^{(1)}(\lambda)$ is given by

$$A(\lambda) = (I_m\lambda - L)\hat{A}^{(1)}(\lambda) \qquad (3.40b)$$

Thus, a complete set of regular right solvents R_k, $k = 1,\ldots,n$ of $A(\lambda)$ is defined as

$$\sigma(R_k) \cap \sigma(R_j) = \phi , \qquad k \neq j \text{ and } j,k = 1,\ldots,n \qquad (3.41a)$$

$$\sigma(A(\lambda)) = \bigcup_{k=1}^{n} \sigma(R_k) \qquad (3.41b)$$

Similarly, a complete set of regular left solvents L_k, $k = 1,\ldots,n$ of $A(\lambda)$ is given by

$$\sigma(L_k) \cap \sigma(L_j) = \phi , \qquad k \neq j, \qquad j,k = 1,\ldots,n \qquad (3.42a)$$

$$\sigma(A(\lambda)) = \bigcup_{k=1}^{n} \sigma(L_k) \qquad (3.42b)$$

An important theorem on regular solvents developed by Markus and Mereuca [23] is:

Theorem 3.14   If $\{R_k,\ k = 1,\ldots,n\}(\{L_k,\ k = 1,\ldots,n\})$ is a complete set of regular right (left) solvents of $A(\lambda)$, then the associated block Vandermonde matrix

$$
V(R_1,\ldots,R_n) \triangleq
\begin{bmatrix}
I_m & I_m & \cdots & I_m \\
R_1 & R_2 & \cdots & R_n \\
R_1^2 & R_2^2 & \cdots & R_n^2 \\
\cdot & \cdot & \cdots & \cdot \\
R_1^{n-1} & R_2^{n-1} & \cdots & R_n^{n-1}
\end{bmatrix}
\tag{3.43a}
$$

or

$$
V^B(L_1,\ldots,L_n) \triangleq
\begin{bmatrix}
I_m & L_1 & L_1^2 & \cdots & L_1^{n-1} \\
I_m & L_2 & L_2^2 & \cdots & L_2^{n-1} \\
\cdot & \cdot & \cdot & \cdots & \cdot \\
I_m & L_n & L_n^2 & \cdots & L_n^{n-1}
\end{bmatrix}
\tag{3.43b}
$$

is nonsingular, where the B in Eq. (3.43b) designates block transpose [18].

From Eq. (3.39b), if $R \epsilon C^{m \times m}$ is a right solvent of a nonsingular λ-matrix $A(\lambda)$, then $A(\lambda)$ can be written as

$$
A(\lambda) = A^{(1)}(\lambda)(I_m \lambda - R)
\tag{3.44a}
$$

where $A^{(1)}(\lambda) \epsilon C^{m \times m}[\lambda]$ is a λ-matrix with degree lower than that of $A(\lambda)$. As we shall see in Chapter IV, $(I_m \lambda - R)$ is a right divisor of $A(\lambda)$, and $A^{(1)}(\lambda)$ is a left divisor of $A(\lambda)$. Similarly from Eq. (3.40b) if $L \epsilon C^{m \times m}$ is a left solvent of $A(\lambda)$, then $A(\lambda)$ can be decomposed as

$$
A(\lambda) = (I_m \lambda - L)\hat{A}^{(1)}(\lambda)
\tag{3.44b}
$$

where $\hat{A}^{(1)}(\lambda) \epsilon C^{m \times m}[\lambda]$ is a λ-matrix with degree lower than that of $A(\lambda)$; and

$(I_m\lambda-L)$ and $\hat{A}^{(1)}(\lambda)$ are a left and a right divisor of $A(\lambda)$, respectively. The more general structure and properties of left and right divisors of a nonsingular λ-matrix are presented in Chapter IV.

3.4 Illustrative Examples

Two examples are presented in this section. The first example demonstrates the relationships between the latent structures of canonical λ-matrices and the eigenstructures of the system map of the associated controller canonical minimal realization. The second example is devoted to verifying the relationships between the latent structures of nonsingular λ-matrices and the latent structures of canonical λ-matrices. In addition the complete set of solvents of nonsingular λ-matrices is found.

Example 3.1

Given a column-reduced canonical λ-matrix

$$D_r(\lambda) = \begin{bmatrix} \lambda^3-3\lambda-2 & 0 & -1 \\ 3\lambda+3 & \lambda^2+2\lambda+1 & -2 \\ 0 & 0 & 1 \end{bmatrix}$$

The controller canonical minimal realization of $D_r^{-1}(\lambda)$ can be found as

$$D_r^{-1}(\lambda) = \bar{C}_c(\lambda I_4-A_c)^{-1}B_c+\bar{D}_c$$

where

$$A_c = \left[\begin{array}{ccc|cc} 0 & 1 & 0 & 0 & 0 \\ 0 & 0 & 1 & 0 & 0 \\ 2 & -3 & 0 & 0 & 0 \\ \hline 0 & 0 & 0 & 0 & 1 \\ -3 & -3 & 0 & -1 & -2 \end{array}\right] ; \quad B_c = \left[\begin{array}{ccc} 0 & 0 & 0 \\ 0 & 0 & 0 \\ \hline 1 & 0 & 1 \\ \hline 0 & 0 & 0 \\ 0 & 1 & 2 \end{array}\right]$$

$$\bar{C}_c = \begin{bmatrix} 1 & 0 & 0 & 0 & 0 \\ 0 & 0 & 0 & 1 & 0 \\ 0 & 0 & 0 & 0 & 0 \end{bmatrix} ; \quad \bar{D}_c = \begin{bmatrix} 0 & 0 & 0 \\ 0 & 0 & 0 \\ 0 & 0 & 1 \end{bmatrix}$$

The modal matrix of A_c can be found as

$$M_A = \begin{bmatrix} 0.0 & -0.33333 & -0.11111 & 0.33333 & 0.11111 \\ 0.0 & 0.33333 & -0.22222 & -0.33333 & 0.22222 \\ 0.0 & -0.33333 & 0.55556 & 0.33333 & 0.44444 \\ 1.0 & 0.33333 & -0.88889 & -0.33333 & -0.11111 \\ -1.0 & 0.66667 & 1.22222 & 0.33333 & -0.22222 \end{bmatrix}$$

The minimal realization of $D_r^{-1}(\lambda)$ in a Jordan form becomes

$$A_J = M_A^{-1} A M_A = \text{block diag} \begin{bmatrix} \begin{pmatrix} -1 & 1 & 0 \\ 0 & -1 & 1 \\ 0 & 0 & -1 \end{pmatrix} , \begin{pmatrix} -1 & 0 \\ 0 & 2 \end{pmatrix} \end{bmatrix}$$

$$B_J = M_A^{-1} B = \begin{bmatrix} 1 & 0 & 1 & 0 & 1 \\ 0 & 1 & 0 & 1 & 0 \\ 1 & 2 & 1 & 2 & 1 \end{bmatrix}^T ; \quad \bar{D}_J = \bar{D} = \begin{bmatrix} 0 & 0 & 0 \\ 0 & 0 & 0 \\ 0 & 0 & 1 \end{bmatrix}$$

$$\bar{C}_J = \bar{C} M_A = \begin{bmatrix} 0.0 & -0.33333 & -0.11111 & 0.33333 & 0.11111 \\ 1.0 & 0.33333 & -0.88889 & -0.33333 & -0.11111 \\ 0.0 & 0.0 & 0.0 & 0.0 & 0.0 \end{bmatrix}$$

The eigenvalues and the lengths of associated Jordan chains are $\{\lambda_1 = -1, \lambda_2 = -1, \lambda_3 = 2\}$ and $\{\ell_1 = 3, \ell_2 = 1, \ell_3 = 1\}$, respectively. From Theorem 3.9 we have the left Jordan chain of $D_r(\lambda)$

$$(P_{10}, P_{11}, P_{12}) = \begin{bmatrix} 1 & 0 & 1 \\ 0 & 1 & 0 \\ 1 & 2 & 1 \end{bmatrix} ; \quad (P_{20}) = \begin{bmatrix} 0 \\ 1 \\ 2 \end{bmatrix} ; \quad (P_{30}) = \begin{bmatrix} 1 \\ 0 \\ 1 \end{bmatrix}$$

Thus, from Theorem 3.8, we have the left generalized eigenvectors of A_c as

$$P_{cij} = T_P \sum_{k=0}^{j} \frac{1}{k!} \psi_r^{(k)}(\lambda_i) D_{rh}^T P_{i(j-k)}$$

where

$$T_p = \begin{bmatrix} -3 & 0 & 1 & 3 & 0 \\ 0 & 1 & 0 & 0 & 0 \\ 1 & 0 & 0 & 0 & 0 \\ 0 & 0 & 0 & 2 & 1 \\ 0 & 0 & 0 & 1 & 0 \end{bmatrix} ; \quad \psi_r^{(1)}(\lambda) = \begin{bmatrix} 0 & 0 & 0 \\ 1 & 0 & 0 \\ 2\lambda & 0 & 0 \\ 0 & 0 & 0 \\ 0 & 1 & 0 \end{bmatrix} ; \quad \psi_r^{(2)}(\lambda) = \begin{bmatrix} 0 & 0 & 0 \\ 0 & 0 & 0 \\ 2 & 0 & 0 \\ 0 & 0 & 0 \\ 0 & 0 & 0 \end{bmatrix}$$

Thus,

$$P_{c10} = T_p \psi_r(\lambda_1) D_{rh}^T P_{10} = [-2,-1,1,0,0]^T$$

$$P_{c11} = T_p [\psi_r(\lambda_1) D_{rh}^T P_{11} + \psi_r^{(1)}(\lambda_1) D_{rh}^T P_{10}] = [1,1,0,1,1]^T$$

$$P_{c12} = T_p [\psi_r(\lambda_1) D_{rh}^T P_{12} + \psi_r^{(1)}(\lambda_1) D_{rh}^T P_{11} + \frac{1}{2} \psi_r^{(2)} D_{rh}^T P_{10}] = [-1,-1,1,1,0]^T$$

$$P_{c20} = T_p \psi_r(\lambda_2) D_{rh}^T P_{20} = [3,0,0,1,1]^T$$

$$P_{c30} = T_p \psi_r(\lambda_3) D_{rh}^T P_{30} = [1,2,1,0,1]^T$$

If the left generalized eigenvectors P_{cij} are known, then the left generalized latent vectors of $D_r(\lambda)$ can be determined from Theorem 3.7 as

$$P_{ij} = B_c^T P_{cij}; \quad 0 \le j \le \ell_i - 1, \quad 1 \le i \le 3 .$$

From $\bar{C}_J$, the right Jordan chains of $D_r(\lambda)$ can be determined as

$$(q_{10}, q_{11}, q_{12}) = \begin{bmatrix} 0 & -0.33333 & -0.11111 \\ 1 & 0.33333 & -0.88889 \\ 0 & 0.0 & 0.0 \end{bmatrix} ; \quad (q_{20}) = \begin{bmatrix} 0.33333 \\ -0.33333 \\ 0.0 \end{bmatrix} ;$$

$$(q_{30}) = \begin{bmatrix} 0.11111 \\ -0.11111 \\ 0.0 \end{bmatrix} .$$

Thus, from Theorem 3.5, we have the right generalized eigenvectors of A_c as

$$q_{c10} = \psi_r(\lambda_1) q_{10} = [0,0,0,1,-1]^T$$

$$q_{c11} = \psi_r(\lambda_1) q_{11} + \psi_r^{(1)}(\lambda_1) q_{10} = [-0.33333, 0.33333, -0.33333, 0.33333, 0.66667]^T$$

$$q_{c12} = \psi_r(\lambda_1)q_{12} + \psi_r^{(1)}(\lambda_1)q_{11} + \frac{1}{2}\psi_r^{(2)}(\lambda_1)q_{10} = [-0.11111, -0.22222, 0.55556,$$
$$-0.88889, 1.22222]^T$$

$$q_{c20} = \psi_r(\lambda_2)q_{20} = [0.33333, -0.33333, 0.33333, -0.33333, 0.33333]^T$$

$$q_{c30} = \psi_r(\lambda_3)q_{30} = [0.11111, 0.22222, 0.44444, -0.11111, -0.22222]^T$$

From Theorem 3.6, we can compute the right generalized latent vectors of $D_r(\lambda)$ if the right generalized eigenvectors of A_c are known. Using the calculated generalized eigenvectors, it can be easily verified that

$$[p_{c12}^T, p_{c11}^T, p_{c10}^T, p_{c20}^T, p_{c30}^T]^T[q_{c10}, q_{c11}, q_{c12}, q_{c20}, q_{c30}] = I_5 .$$

Example 3.2

Consider a nonsingular λ-matrix

$$D(\lambda) = \begin{bmatrix} \lambda^4 + 9\lambda^3 + 31\lambda^2 + 50\lambda + 33 & \lambda^4 + 10\lambda^3 + 38\lambda^2 + 67\lambda + 49 \\ \lambda^2 + 7\lambda + 12 & \lambda^2 + 8\lambda + 18 \end{bmatrix}$$

which is not a canonical λ-matrix (check with Definition 3.1). Using column equivalent transformation (Theorem 3.3), we have

$$D(\lambda) = D_r(\lambda)U_r(\lambda)$$

where

$$D_r(\lambda) = \begin{bmatrix} \lambda^3 + 7\lambda^2 + 17\lambda + 17 & -1 \\ 6 & \lambda \end{bmatrix}$$

and

$$U_r(\lambda) = \begin{bmatrix} \lambda+2 & \lambda+3 \\ \lambda+1 & \lambda+2 \end{bmatrix}, \quad U_r^{-1}(\lambda) = \begin{bmatrix} \lambda+2 & -\lambda-3 \\ -\lambda-1 & \lambda+2 \end{bmatrix}$$

It is easy to verify that $D_r(\lambda)$ is a column-reduced canonical λ-matrix and $U_r(\lambda)$ is a unimodular λ-matrix. Following the same approach in Example 3.1, we have the Jordan form minimal realization of $D_r^{-1}(\lambda)$:

$$D_r^{-1}(\lambda) = \overline{C}_J(\lambda I_r - A_J)^{-1} B_J + D_J$$

where

$$A_J = \begin{bmatrix} -1 & 1 & 0 & 0 \\ 0 & -1 & 0 & 0 \\ 0 & 0 & -3 & 0 \\ 0 & 0 & 0 & -2 \end{bmatrix} \qquad B_J = \begin{bmatrix} 1 & 0 \\ -1 & 1 \\ 3 & -1 \\ 2 & -1 \end{bmatrix}$$

$$\overline{C}_J = \begin{bmatrix} 0.5 & -0.75 & 0.25 & -1.0 \\ 3.0 & -1.5 & 0.5 & -3.0 \end{bmatrix}$$

$$\overline{D}_J = \begin{bmatrix} 0 & 0 \\ 0 & 0 \end{bmatrix}$$

The latent roots and the lengths of associated Jordan chains of $D_r(\lambda)$ are $\{\lambda_1 = -1, \lambda_2 = -3, \lambda_3 = -2\}$ and $\{\ell_1 = 2, \ell_2 = 1, \ell_3 = 1\}$, respectively. From Theorem 3.9, we have the left Jordan chains of $D_r(\lambda)$:

$$(P_{10}, P_{11}) = \begin{bmatrix} 1 & -1 \\ 0 & 1 \end{bmatrix}, \quad (P_{20}) = \begin{bmatrix} 3 \\ -1 \end{bmatrix}, \quad (P_{30}) = \begin{bmatrix} 2 \\ -1 \end{bmatrix}$$

and the right Jordan chains:

$$(q_{10}, q_{11}) = \begin{bmatrix} 0.5 & -0.75 \\ 3.0 & -1.5 \end{bmatrix}, \quad (q_{20}) = \begin{bmatrix} 0.25 \\ 0.5 \end{bmatrix}, \quad (q_{30}) = \begin{bmatrix} -1.0 \\ -3.0 \end{bmatrix}$$

Using Theorem 3.11, we have a left Jordan chains of $D(\lambda)$:

$$(\bar{p}_{10}, \bar{p}_{11}) = \begin{bmatrix} 1 & -1 \\ 0 & 1 \end{bmatrix}, \quad (\bar{p}_{20}) = \begin{bmatrix} 3 \\ -1 \end{bmatrix}, \quad (\bar{p}_{30}) = \begin{bmatrix} 2 \\ -1 \end{bmatrix}$$

and the right Jordan chains:

$$\bar{q}_{10} = U_r^{-1}(\lambda_1) q_{10} = \begin{bmatrix} 1 & -2 \\ 0 & 1 \end{bmatrix} \begin{bmatrix} 0.5 \\ 3.0 \end{bmatrix} = \begin{bmatrix} -5.5 \\ 3.0 \end{bmatrix}$$

$$\bar{q}_{11} = U_r^{-1}(\lambda_1) q_{11} + U_r^{-(1)} q_{10}$$

$$= \begin{bmatrix} 1 & -2 \\ 0 & 1 \end{bmatrix} \begin{bmatrix} -0.75 \\ -1.5 \end{bmatrix} + \begin{bmatrix} 1 & -1 \\ -1 & 1 \end{bmatrix} \begin{bmatrix} 0.5 \\ 3.0 \end{bmatrix} = \begin{bmatrix} -0.25 \\ 1.0 \end{bmatrix}$$

$$\bar{q}_{20} = U_r^{-1}(\lambda_2) q_{20} = \begin{bmatrix} -1 & 0 \\ 2 & -1 \end{bmatrix} \begin{bmatrix} 0.25 \\ 0.5 \end{bmatrix} = \begin{bmatrix} -0.25 \\ 0 \end{bmatrix}$$

$$\bar{q}_{30} = U_r^{-1}(\lambda_3) q_{30} = \begin{bmatrix} 0 & -1 \\ 1 & 0 \end{bmatrix} \begin{bmatrix} -1.0 \\ -3.0 \end{bmatrix} = \begin{bmatrix} 3.0 \\ -1.0 \end{bmatrix}$$

From the left Jordan chains of $D(\lambda)$ and Theorem 3.13, we have a complete set of left solvents:

$$L_1 = [\bar{p}_{10} \ \bar{p}_{11}]^{-T} \begin{bmatrix} \lambda_1 & 1 \\ 0 & \lambda_1 \end{bmatrix} [\bar{p}_{10} \ \bar{p}_{11}]^T$$

$$= \begin{bmatrix} 1 & 0 \\ 1 & 1 \end{bmatrix} \begin{bmatrix} -1 & 1 \\ 0 & -1 \end{bmatrix} \begin{bmatrix} 1 & 0 \\ -1 & 1 \end{bmatrix} = \begin{bmatrix} -2 & 1 \\ -1 & 0 \end{bmatrix}$$

$$L_2 = [\bar{P}_{20} \ \bar{P}_{30}]^{-T} \begin{bmatrix} \lambda_2 & 0 \\ 0^2 & \lambda_3 \end{bmatrix} [\bar{P}_{20} \ \bar{P}_{30}]^{T}$$

$$= \begin{bmatrix} 1 & -1 \\ 2 & -3 \end{bmatrix} \begin{bmatrix} -3 & 0 \\ 0 & -2 \end{bmatrix} \begin{bmatrix} 3 & -1 \\ 2 & -1 \end{bmatrix} = \begin{bmatrix} -5 & 1 \\ -6 & 0 \end{bmatrix}$$

Similarly, from the right Jordan chains of $D(\lambda)$ and Theorem 3.13, we have a complete set of right solvents:

$$R_1 = [\bar{q}_{10} \ \bar{q}_{11}] \begin{bmatrix} \lambda_1 & 1 \\ 0 & \lambda_1 \end{bmatrix} [\bar{q}_{10} \ \bar{q}_{11}]^{-1}$$

$$= \begin{bmatrix} -5.5 & -0.25 \\ 3.0 & 1.0 \end{bmatrix} \begin{bmatrix} -1 & 1 \\ 0 & -1 \end{bmatrix} \begin{bmatrix} 1.0 & 0.25 \\ -3.0 & -5.5 \end{bmatrix} (\frac{-1}{4.75})$$

$$= \frac{1}{19} \begin{bmatrix} -85 & -121 \\ 36 & 47 \end{bmatrix}$$

$$R_2 = [\bar{q}_{20} \ \bar{q}_{30}] \begin{bmatrix} \lambda_2 & 0 \\ 0 & \lambda_3 \end{bmatrix} [\bar{q}_{20} \ \bar{q}_{30}]^{-1}$$

$$= \begin{bmatrix} -0.25 & 3.0 \\ 0 & -1.0 \end{bmatrix} \begin{bmatrix} -3 & 0 \\ 0 & -2 \end{bmatrix} \begin{bmatrix} -4 & -12 \\ 0 & -1 \end{bmatrix}$$

$$= \begin{bmatrix} -3 & -3 \\ 0 & -2 \end{bmatrix} .$$

CHAPTER IV   DIVISORS AND SPECTRAL FACTORS OF NONSINGULAR λ-MATRICES

An essential difficulty in structural decompositions [30,31] of MIMO systems lies in determining the left/right divisors and spectral factors of a nonsingular λ-matrix. The state-space minimal realizations of the inverse of a nonsingular λ-matrix discussed in Chapter III will be used to determine the left/right divisors [30-31] and spectral factors [51-53]. In this chapter, we investigate the existence of the left/right divisors and the spectral factors of a nonsingular λ-matrix. The so-called geometric approaches [5,47-50,81-82] are employed to derive the structure decomposition theorems of a nonsingular λ-matrix.

First, we shall define the left and right divisors of λ-matrices as follows:

Definition 4.1   Given $A(\lambda) \epsilon C^{m \times m}[\lambda]$. If

$$A(\lambda) = L(\lambda)R(\lambda) \tag{4.1}$$

where $L(\lambda)$, $R(\lambda) \epsilon C^{m \times m}[\lambda]$, then $L(\lambda)$ is a left divisor and $R(\lambda)$ is a right divisor of $A(\lambda)$, and $L(\lambda)$, $R(\lambda)$ are left and right spectral factors of $A(\lambda)$. If $det(L(\lambda)) \neq 0$, $(det(R(\lambda)) \neq 0)$, then $L(\lambda)$, $(R(\lambda))$, is a nonsingular left (right) divisor of $A(\lambda)$. Also, if $L(\lambda)$, $(R(\lambda))$, is a column-reduced (row-reduced) canonical λ-matrix, then $L(\lambda)$, $(R(\lambda))$ is a canonical left (right) divisor of $A(\lambda)$.   □

Definition 4.2   A left (right) divisor $L(\lambda)$ $(R(\lambda))$ of $A(\lambda)$ is nontrivial if $L(\lambda)$ $(R(\lambda))$ is not a unimodular λ-matrix.   □

If $U(\lambda)$ is a unimodular λ-matrix, $L(\lambda)$, $(R(\lambda))$, is a left (right) divisor of $A(\lambda)$, then $L(\lambda)U(\lambda)$, $(U(\lambda)R(\lambda))$, is also a left (right) divisor of $A(\lambda)$.

Furthermore, from Theorem 3.3 any nonsingular λ-matrix can be reduced to a column (row) reduced canonical λ-matrix by post multiplying (premultiplying) a unimodular λ-matrix. Thus, to investigate the structures of left/right divisors of a nonsingular λ-matrix, it is sufficient to study the structures of canonical left (right) divisors of a column (row) reduced canonical λ-matrix.

In Section 4.1, we shall discuss the state-space structures of the canonical left divisors of column-reduced canonical λ-matrices; the state-space structures of the canonical right divisors of row-reduced canonical λ-matrices are dealt with in Section 4.2. Also, constructive proofs on the existence and properties of canonical left/right divisors and complete sets of canonical left/right divisors are provided in Sections 4.1 and 4.2. Section 4.3 is devoted to investigating the state-space structures of the spectral factorization of nonsingular λ-matrices. The computational aspects of the left/right canonical divisors are presented in Section 4.4. The main theorems used to implement the computational algorithms are block diagonalization and block triangularization [54,55] of the system map, which is the minimal realization quadruples of the inverse of a nonsingular λ-matrix. A numerical method to compute the left/right canonical divisors using the matrix sign algorithm [56-60] is presented. Some illustrative examples are provided in Section 4.5.

4.1 Structure Theorems for Canonical Left Divisors and Complete Sets of Canonical Left Divisors

In this section, we develop the structure theorems for canonical left divisors of a column-reduced canonical λ-matrix or a right characteristic λ-matrix of a reachable pair (A,B) in an MIMO system. The main tool used to study the structural theorems of this section is the reachable state-space realization [5,47-50] for the inverse of column-reduced canonical λ-matrices. The complete set of canonical left divisors of a column-reduced canonical λ-matrices is defined, together with the structure theorem for complete sets of canonical left divisors.

Consider a reachable pair (A,B). Let $X = C^n$ be the associated state space, $U = C^m$ be the input space, and A: $X \to X$, B: $U \to X$. Also, let $S = C^{n_s}$ be an A-invariance subspace of X where $n_s < n$, or $AS \subset S$, and assume that V is the canonical projection of X on X/S, or V: $X \to X/S$. Denote $B = \text{Im}(B)$, and $B_L = (B+S)/S$. Let A_L be the induced map in X/S by A, and $B_L = VB$, then $B_L = \text{Im}(B_L)$ and the following diagram commutes:

$$
\begin{array}{ccc}
& X & \xrightarrow{\ A\ } & X \\
{\scriptstyle B}\nearrow & \downarrow V & & \downarrow V \\
U & & & \\
{\scriptstyle B_L}\searrow & X/S & \xrightarrow[A_L]{} & X/S
\end{array}
\qquad (4.2)
$$

It is well known that V is epic [5], and if (A,B) is reachable, then the induced pair (A_L, B_L) in X/S is also reachable, or $X/S = \langle A_L | B_L \rangle \overset{\Delta}{=} B_L + A_L B_L + \ldots + A_L^{n_v - 1} B_L$ which is the reachability subspace [5] of (A_L, B_L) and $n_v = n - n_s$.

Theorem 4.1   Let (A,B) and (A_L, B_L), where $A \epsilon C^{n \times n}$, $B \epsilon C^{n \times m}$, $A_L \epsilon C^{n_v \times n_v}$, $B_L \epsilon C^{n_v \times m}$, be reachable pairs with right characteristic λ-matrices $D_r(\lambda)$ and $D_{LR}(\lambda)$, respectively, and $n \geq n_v > 0$. Then $D_{Lr}(\lambda)$ is a canonical left divisor of $D_r(\lambda)$ iff (A_L, B_L) is an induced pair of (A,B).

Proof:

(i) Sufficient Part:

Let X be the state space of (A,B), S be an A-invariant subspace of X and V: $X \to X/S$ be a canonical projection. Then, from Eq. (4.2), we have

$$A_L V = VA \qquad (4.3a)$$

$$B_L = VB \qquad (4.3b)$$

Also, from Eq. (2.14b), we obtain

$$BD_r(\lambda) = (\lambda I_n - A)T_c^{-1}\psi_r(\lambda) \ .$$ (4.4a)

Premultiplying Eq. (4.4a) by V yields

$$VBD_r(\lambda) = V(\lambda I_n - A)T_c^{-1}\psi_r(\lambda)$$

or

$$B_L D_r(\lambda) = (\lambda I_{n_v} - A_L)VT_c^{-1}\psi_r(\lambda)$$ (4.4b)

Since (A_L, B_L) is an induced pair of (A,B), (A_L, B_L) is reachable and the inverse of the right characteristic λ-matrix of (A_L, B_L) can be realized using Lemma 2.1 as

$$D_{Lr}^{-1}(\lambda) = \overline{C}_L(\lambda I_{n_v} - A_L)^{-1}B_L + \overline{D}_L$$ (4.4c)

where

$$\overline{C}_L \triangleq \psi_{LR}^T(0)T_{LC}; \ \overline{D}_L = [I_m - \psi_{LR}^T(0)\psi_{LR}(0)]D_{Lrh}^{-1}; \ \psi_{Lr}(\lambda), T_{LC}, D_{Lrh}$$

are defined as in Eqs. (2.14c), (2.11b) and (2.9a) for the reachable pair (A_L, B_L). Using the results in Eqs. (4.4b) and (4.4c) yields

$$D_{Lr}^{-1}(\lambda)D_r(\lambda) = \overline{C}_L VT_c^{-1}\psi_r(\lambda) + \overline{D}_L D_r(\lambda)$$ (4.5)

Since the right-hand side of Eq. (4.5) is a λ-matrix, $D_{Lr}(\lambda)$ is a column reduced left divisor of $D_r(\lambda)$.

(2) Necessary Part

Let $D_{Lr}(\lambda)R(\lambda) = D_r(\lambda)$ where $R(\lambda)\epsilon C^{m \times m}[\lambda]$. Since

$$B_L D_{Lr}(\lambda) = (\lambda I_{n_v} - A_L) T_{LC}^{-1} \psi_{LR}(\lambda), \text{ we have}$$

$$B_L D_r(\lambda) = (\lambda I_{n_v} - A_L) T_{LC}^{-1} \psi_{Lr}(\lambda) R(\lambda) \ . \tag{4.6}$$

Thus $\partial_{ci}(D_r(\lambda)) > \partial_{ci}(\psi_{Lr}(\lambda) R(\lambda))$, and $\psi_{Lr}(\lambda) R(\lambda)$ can be written as

$$\psi_{Lr}(\lambda) R(\lambda) = W \psi_r(\lambda), \ W \epsilon C^{n_v \times n} \ . \tag{4.7}$$

Substituting Eq. (4.7) into Eq. (4.6) gives

$$B_L D_r(\lambda) = (\lambda I_{n_v} - A_L) T_{LC}^{-1} W \psi_r(\lambda)$$

or

$$(\lambda I_{n_v} - A_L)^{-1} B_L = T_{LC}^{-1} W \psi_r(\lambda) D_r^{-1}(\lambda) \ . \tag{4.8a}$$

Using the results in Eqs. (2.13), (2.14b) and (2.15b), Eq. (4.8a) becomes

$$(\lambda I_{n_v} - A_L)^{-1} B_L = T_{LC}^{-1} W [(\lambda I_n - A) T_C^{-1}]^{-1} B \tag{4.8b}$$

Define $V \overset{\Delta}{=} T_{LC}^{-1} W T_C \epsilon C^{n_v \times n}$. Then Eq. (4.8b) can be expressed as

$$(\lambda I_{n_v} - A_L)^{-1} B_L = V(\lambda I_n - A)^{-1} B$$

or

$$\sum_{i=1}^{\infty} A_L^{i-1} B_L \lambda^{-i} = \sum_{i=1}^{\infty} V A^{i-1} B \lambda^{-i} \tag{4.9}$$

By equating the matrix coefficients in Eq. (4.9), we get

$$A_L^i B_L = VA^i B, \quad i \geq 0$$

or

$$[A_L V - VA][B, AB, A^2 B, \ldots] = [0_{n_v \times m}, 0_{n_v \times m}, \ldots]$$

Since (A, B) is reachable, we have

$$A_L V = VA; B_L = VB \tag{4.10}$$

Also, since (A_L, B_L) is reachable and

$$[B_L, A_L B_L, \ldots, A_L^{n_v - 1} B_L] = V[B, AB, \ldots, A^{n_v - 1} B]$$

we have

$$\text{rank}[B_L, A_L B_L, \ldots, A_L^{n_v - 1} B_L] = n_v \leq \min[\text{rank}(V), \text{rank}(B, AB, \ldots, A^{n_v - 1} B)] .$$

Thus, $\text{rank}(V) \geq n_v$. But $V \in C^{n_v \times n}$ so we obtain $\text{rank}(V) = n_v$, ie. V is epic. Thus, V is a canonical projection $V: X \rightarrow X/S$, where $S = \text{Ker}(V)$ and S is an A-invariant subspace of X. From the results in Eq. (4.10) we can conclude that (A_L, B_L) is an induced pair of (A, B) in X/S via V. ∎

Some structure properties of the left divisors of $D_r(\lambda)$ are described as follows.

<u>Lemma 4.1</u>   Let $(\kappa_1, \kappa_2, \ldots, \kappa_m)$ be the Kronecker indices of (A, B) and $(\kappa_{L1}, \kappa_{L2}, \ldots, \kappa_{Lm})$ be the Kronecker indices of (A_L, B_L), then

$$\kappa_{Li} \leq \kappa_i , \quad 1 \leq i \leq m$$

<u>Proof:</u>

Let $B \overset{\Delta}{=} [b_1,\ldots,b_m]$, $B_L \overset{\Delta}{=} [b_{L1},\ldots,b_{Lm}]$

(1) Let $i=1$.

Assume, by contradiction, that $\kappa_{L1} > \kappa_1$. Then, we obtain $\{b_{L1}, A_L b_{L1}, \ldots,$ $A_L^{\kappa_{L1}-1} b_{L1}\} = \{V b_1, V A B_1, \ldots, V A^{\kappa_{L1}-1} b_1\}$ which contains independent vectors. However, $A^k b_1$ for $k > \kappa_i$ is dependent of $A^j b_1$ for $j = 1,\ldots,\kappa_i$. Thus, $V A^k b_1 (= A_L^k b_{L1})$ for $k > \kappa_i$ is dependent of $V A^j b_1 (= A_L^j b_{L1})$ for $j = 1,\ldots,\kappa_i$. The above result contradicts to the assumption, and therefore $\kappa_{L1} < \kappa_1$.

(2) Let $\kappa_{Li} \leq \kappa_i$ for $i \leq i_k - 1$.

Assume, by contradiction, that $\kappa_{Li_k} > \kappa_{i_k}$. Then, we obtain

$$\{b_{L1}, A_L b_{L1}, \ldots, A_L^{\kappa_{L1}-1} b_{L1}, \ldots, b_{Li_k}, A_L b_{Li_k}, \ldots, A_L^{\kappa_{Li_k}-1} b_{Li_k}\} =$$

$$\{V b_1, V A b_1, \ldots, V A^{\kappa_{L1}-1} b_1, \ldots, V b_{i_k}, V A b_{i_k}, \ldots, V A^{\kappa_{Li_k}-1} b_{i_k}\} \quad \text{which} \quad \text{contains}$$

independent vectors. However, for $k > \kappa_{i_k}$.

$$A^k b_{i_k} = \sum_{\xi=1}^{i_k} \sum_{j=1}^{\kappa_\xi-1} \alpha_{k\xi j} A^j b_\xi .$$

So that

$$V A^k b_{i_k} = \sum_{\xi=1}^{i_k} \sum_{j=1}^{\kappa_\xi-1} \alpha_{k\xi j} V A^j b_\xi$$

From the hypothesis $\kappa_{Li} \leq \kappa_i$ for $i \leq i_k$, we have

$$\sum_{j=1}^{\kappa_\xi-1} \alpha_{k\xi j} V A^j b_\xi = \sum_{j=1}^{\kappa_{L\xi}-1} \bar{\alpha}_{k\xi j} V A^j b_\xi$$

which implies, $V A^k b_{i_k} = A_L^k b_{Li_k} = \sum_{\xi=1}^{i_k} \sum_{j=1}^{\kappa_{L\xi}-1} \bar{\alpha}_{k\xi j} A_L^j b_{L\xi} .$

This leads to a contradiction, and therefore $\kappa_{Li_k} \le \kappa_{i_k}$.

(3) By mathematical induction, we obtain

$$\kappa_{Li} \le \kappa_i, \qquad 1 \le i \le m \qquad\blacksquare$$

Lemma 4.2 Let $P(A,B)$ and $P(A_L,B_L)$ be the reachability base matrices of (A,B) and (A_L,B_L), and $(\kappa_1,\ldots,\kappa_m)$, let $(\kappa_{L1},\ldots,\kappa_{Lm})$ be the Kronecker indices of (A,B) and (A_L,B_L), respectively and let V be the canonical projection shown in Eq. (4.2). Then

$$V P(A,B) H = P(A_L,B_L) \qquad\qquad (4.11)$$

where

$$H = [e_n^1,\ldots,e_n^{\kappa_{L1}},e_n^{\sigma_1+1},\ldots,e_n^{\sigma_1+\kappa_{L2}},\ldots,e_n^{\sigma_{m-1}+1},\ldots,e_n^{\sigma_{m-1}+\kappa_{Lm}}] \in C^{n \times n_v}$$

$$\sigma_i = \sum_{j=1}^{i} \kappa_j, \qquad 1 \le i \le m$$

Proof:

Since $V P(A,B) = [b_{L1},A_L b_{L1},\ldots,A_L^{\kappa_1-1} b_{L1},\ldots,b_{Lm},A_L b_{Lm},\ldots,A_L^{\kappa_m-1} b_{Lm}]$, from Lemma 4.1 and the definition of H, the result in Lemma 4.2 follows.

Theorem 4.2 Let V, H, $P(A,B)$, $P(A_L,B_L)$ be the same as those defined in Lemma 4.2. Then

$$V = P(A_L,B_L) H^T P(A,B)^{-1} \qquad\qquad (4.12)$$

Proof:

The result follows from Lemma 4.2 and the fact that $H^T H = I_{n_v}$. $\blacksquare$

<u>Lemma 4.3</u>   Let T_c be the transformation matrix for the controller canonical form.   Then $T_c = T_{cp}^{-1} P^{-1}(A,B)$ where

$$T_{cp} = [T_{cp}]_{ij}; \quad \kappa_i \kappa_j > 0; \quad 1 \leq i \leq m; \quad 1 \leq j \leq m$$

$[T_{cp}]_{ii}$ =   The reversed upper triangular Toeplitz matrix with first row

$[a_{rii2}, \ldots, a_{rii\kappa_i}, 1]$

$[T_{cp}]_{ij}$ =   The reversed upper triangular Toeplitz matrix with first row

$[a_{rij2}, \ldots, a_{rij\kappa_j}, 0]$ if $\kappa_i \leq \kappa_j$ or $[a_{rij2}, \ldots, a_{rij\kappa_i}, 0, \ldots, 0]$ if $\kappa_i < \kappa_j$.

<u>Proof:</u>

Lemma 4.3 can be verified by direct computations.   ∎

<u>Corollary 4.1</u>   Let T_c and T_{LC} be the similarity transformation matrices which transform (A,B) and (A_L, B_L) to their controller forms, then

$$V = T_{LC}^{-1} W T_C \qquad\qquad (4.13)$$

where

$$W = T_{Lcp}^{-1} H^T T_{cp}; \quad T_{Lcp}^{-1} = T_{LC} P(A_L, B_L); \quad T_{cp} = P^{-1}(A,B) T_C^{-1}$$

<u>Proof:</u>

From Lemma 4.3, we obtain $P(A,B) = T_c^{-1} T_{cp}^{-1}$ and $P(A_L, B_L) = T_{LC}^{-1} T_{Lcp}^{-1}$.   Thus, from Theorem 4.2 we have

$$V = P(A_L, B_L) H^T P^{-1}(A,B) = T_{LC}^{-1} T_{Lcp}^{-1} H^T T_{cp} T_C = T_{LC}^{-1} W T_C$$

where $W = T_{LCP}^{-1} H^T T_{cp}$. ∎

From Eq. (4.5), the properties of the right divisor pairing with the canonical left divisor $D_{Lr}(\lambda)$ of $D_r(\lambda)$ can be stated as follows.

Corollary 4.2   Let $D_{Lr}(\lambda)$ be a canonical left divisor of $D_r(\lambda)$, which is defined in Theorem 4.1, and $D_r(\lambda) = D_{Lr}(\lambda)R(\lambda)$. Then

(1) $R(\lambda) = \psi_{Lr}^T(0)W\psi_r(\lambda) + (I_m - \psi_{Lr}^T(0)\psi_{Lr}(0))D_{Rh}\delta_r(\lambda)$

where $D_{Rh} = D_{Lrh}^{-1}D_{rh}$ is an upper triangular matrix with diagonal elements all 1's.

(2) $R(\lambda)$ is a nonsingular λ-matrix.

Proof:

(1) $R(\lambda) = \overline{C}_L V T_C^{-1}\psi_r(\lambda) + \overline{D}_L D_r(\lambda)$

$\qquad = \psi_{Lr}^T(0)T_{LC}T_{LC}^{-1}WT_C T_c^{-1}\psi_r(\lambda) + (I_m - \psi_{Lr}^T(0)\psi_{Lr}(0))D_{Lrh}^{-1}D_{rh}\delta_r(\lambda)$

$\qquad = \psi_{Lr}^T(0)W\psi_r(\lambda) + (I_m - \psi_{Lr}^T(0)\psi_{Lr}(0))D_{Rh}\delta_r(\lambda)$

(2)   Since both $D_r(\lambda)$ and $D_{Lr}(\lambda)$ are column-reduced canonical λ-matrices, $R(\lambda)$ is nonsingular. ∎

For some applications (refer to Chapter VI) of the divisor theorems, we need to have a full set of divisors whose spectra cover the entire spectrum of a λ-matrix. We shall discuss the complete set of left divisors before leaving this section.

<u>Definition 4.3</u>   If $L_c = \{D_i(\lambda), i=1,\ldots,k\}$ is a set of left divisors of a nonsingular λ-matrix $A(\lambda)$ such that $D_i(\lambda)$ are nontrivial and $\bigcup_{i=1}^{k} \sigma(D_i(\lambda)) = \sigma(A(\lambda))$, then L_c is a <u>complete set</u> of left divisors of $A(\lambda)$; if $D_i(\lambda)$ are canonical left divisors, then L_c is a complete set of canonical left divisors. If any complete set of left divisors of $D_i(\lambda)$ contains only one element which is column equivalent to $D_i(\lambda)$, then L_c is an <u>irreducible</u> complete set of left divisors of $A(\lambda)$; if $D_i(\lambda)$ are canonical left divisors, then L_c is an irreducible complete set of canonical left divisors.   The complete set of canonical right divisors and irreducible complete set of canonical right divisors of $A(\lambda)$ can be defined in the same manner.   ◻

The structure theorems for a complete set of canonical left divisors are formulated in Theorems 4.3 and 4.4; while the irreducible complete set of left divisors is discussed in Section 4.4.

<u>Theorem 4.3</u>   Let S_1 and S_2 be complemented A-invariant subspaces of X (where $S_1 \cap S_2 = \phi$ and $X = S_1 \oplus S_2$), and (A_{Li}, B_{Li}), $i=1,2$, be the induced reachable pairs of (A,B) in X/S_i, $i=1,2$. Then the following diagram commutes.

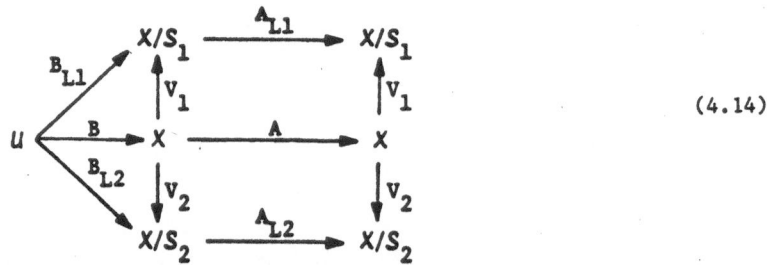

$$(4.14)$$

The set of the right characteristic λ-matrices $\{D_{Lr_i}(\lambda), i=1,2\}$, of (A_{Li}, B_{Li}), $i=1,2$, is a complete set of canonical left divisors of $D_r(\lambda)$.

<u>Proof:</u>

From Theorem 4.1, $D_{Lr_i}, i=1,2$ are canonical left divisors of $D_r(\lambda)$.   Since $V_iA = A_{Li}V_i$, we obtain

$$\begin{bmatrix} V_1 \\ V_2 \end{bmatrix} A = \begin{bmatrix} A_{L1} & 0 \\ 0 & A_{L2} \end{bmatrix} \begin{bmatrix} V_1 \\ V_2 \end{bmatrix} \qquad\qquad (4.15)$$

Because $\text{Ker}(V_i) = S_i$ and $S_1 \cap S_2 = \phi$, we have $\text{Ker}(V_1) \cap \text{Ker}(V_2) = \phi$. Also, $S_1 \oplus S_2 = X$, therefore $\dim(\text{Ker}(V_1)) + \dim(\text{Ker}(V_2)) = \dim(X) = n$. Let $\overline{V} \overset{\Delta}{=} [V_1^T, V_2^T]^T$. Since $\text{Ker}(\overline{V}) = \text{Ker}(V_1) \cap \text{Ker}(V_2) = \phi$, we have $\dim(\overline{V}) = n$, ie. $\overline{V}$ is nonsingular. As a result, we obtain take over $\overline{V} A \overline{V}^{-1} = \text{block diag}(A_{L1}, A_{L2})$, and $\sigma(D_{Lr_1}(\lambda)) \cup \sigma(D_{Lr_2}(\lambda)) = \sigma(A_{L1}) \cup \sigma(A_{L2}) = \sigma(\text{block diag}(A_{L1}, A_{L2})) = \sigma(A) = \sigma(D_r(\lambda))$. Thus, $\{D_{Lr_i}(\lambda), i=1,2\}$ is a complete set of canonical left divisors of $D_r(\lambda)$. ∎

The results of Theorem 4.3 can be easily extended to a more general case as follows.

Theorem 4.4   Let $S_i, i=1,\ldots,k$ be independent A-invariant subspaces of X such that $\overset{k}{\underset{i=1}{\oplus}} S_i = X$. Define $X_i \overset{\Delta}{=} \overset{k}{\underset{\substack{j=1 \\ j \neq i}}{\oplus}} S_i$ and $(A_{Li}, B_{Li}), i=1,\ldots,k$ be the induced reachable pairs of (A,B) in X/X_i, and $D_{Lr_i}(\lambda)$, $i=1,\ldots,k$ be the right characteristic λ-matrices of $(A_{Li}, B_{Li}), i=1,\ldots,k$. Then $\{D_{Lr_i}(\lambda), i=1,\ldots,k\}$ is a complete set of canonical left divisors of $D_r(\lambda)$. ∎

4.2   Structure Theorems for Canonical Right Divisors and Complete Sets of Canonical Right Divisors

Parallel to the structural analysis of the canonical left divisors for column-reduced λ-matrices in Section 4.1, we derive the structure theorems for the canonical right divisors and the complete set of canonical right divisors for row-reduced λ-matrices in this section.

Let $X = C^n$ be the state space, $Y = C^p$ be the output space, and A: $X \rightarrow X$, C: $X \rightarrow Y$. Let $S = C^{n_s}$, $n_s < n$ be an A-invariant subspace of X, and S be the canonical injection map S: $S \rightarrow X$. Define $A_R = A|S$ and $C_R = C|S$. Then the following diagram commutes.

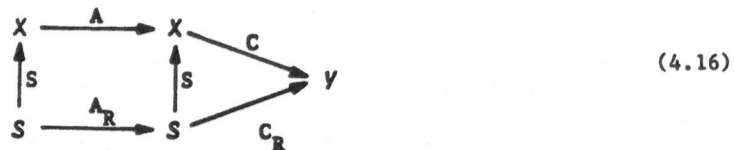

(4.16)

It is known that S is monic, and if (A,C) is observable, then the embedded pair (A_R, C_R) is also observable, or $\bigcap_{i=1}^{n_s} Ker(C_R A_R^{i-1}) = 0$.

Dual to Theorem 4.1, we have the following results.

Theorem 4.5   Let (A,C) and (A_R, C_R) be observable pairs with left characteristic λ-matrices $D_\ell(\lambda)$ and $D_{R\ell}(\lambda)$, respectively and $A \varepsilon C^{n \times n}, C \varepsilon C^{p \times n}, A_R \varepsilon C^{n_s \times n_s}, C_R \varepsilon C^{p \times n_s}, n \geq n_s > 0$. Then $D_{R\ell}(\lambda)$ is a canonical right divisor of $D_\ell(\lambda)$ iff (A_R, C_R) is an embedded pair of (A,C), or

$$D_\ell(\lambda) = L(\lambda) D_{R\ell}(\lambda)$$

where $D_{R\ell}^{-1}(\lambda) = C_R(\lambda I_{n_s} - A_R)^{-1} \bar{B}_R + \bar{D}_R$; $L(\lambda) = \psi_\ell(\lambda) T_0^{-1} S \bar{B}_R + D_\ell(\lambda) \bar{D}_R$; $T_0, \psi_\ell(\lambda)$, and S are defined in Eqs. (2.17b), (2.20c) and (4.16), respectively. ∎

The structure properties of the right divisors of $D_\ell(\lambda)$ are as follows, beginning with the analogues of Lemmas 4.1 and 4.2.

Lemma 4.4   Let $(\nu_1, \nu_2, \ldots, \nu_p)$ be the observability indices of (A,C) and $(\nu_{R1}, \nu_{R2}, \ldots, \nu_{Rp})$ be the observability indices of (A_R, C_R). Then $\nu_{Ri} \leq \nu_i$, $1 \leq i \leq p$. ∎

Lemma 4.5   Let $\mathcal{Q}(A,C)$ and $\mathcal{Q}(A_R, C_R)$ be the observable base matrices of (A,C) and (A_R, C_R), respectively. Let S be the canonical injection map in Eq. (4.16). Then

$$\hat{H} \mathcal{Q}(A,C) S = \mathcal{Q}(A_R, C_R) \tag{4.17}$$

where

$$\hat{H} = [e_n^1,\ldots,e_n^{\nu_{R1}},e_n^{\tau_1+1},\ldots,e_n^{\tau_1+\nu_{R2}},\ldots,e_n^{\tau_{p-1}+1},\ldots,e_n^{\tau_{p-1}+\nu_{Rp}}]^T \in C_s^{n_s \times n}$$

$$\tau_i = \sum_{j=1}^{i} \nu_j \ , \ 1 \le i \le p$$

$(\nu_1,\ldots,\nu_p)$ and $(\nu_{R1},\ldots,\nu_{Rp})$ are the observability indices of (A,C) and (A_R,C_R), respectively. ∎

It then follows directly from Lemma 4.5:

Theorem 4.6   Let S, $\hat{H}$, $Q(A,C)$, and $Q(A_R,C_R)$ be the same as those defined in Lemma 4.4. Then

$$S = Q^{-1}(A,C)\hat{H}^T Q(A_R,C_R) \tag{4.18}$$

∎

Lemma 4.6   Let T_0 be the similarity transformation matrix for the observer canonical form and $Q(A,C)$ be the observability base matrix. Then

$$T_0 = Q^{-1}(A,C)T_{0p}^{-1} \tag{4.19}$$

where

$$T_{0p} = [T_{0p}]_{ij} \in C^{\nu_i \times \nu_j}, \ 1 \le i \le p, \ 1 \le j \le p, \ \nu_i\nu_j > 0$$

$[T_{0p}]_{ii}$ =   The reversed upper triangular Toeplitz matrix with first row $[a_{\ell ii2},\ldots,a_{\ell ii\nu_i},1]$

$[T_{0p}]_{ij}$ =   The reversed upper triangular Toeplitz matrix with first row $[a_{\ell ij2},\ldots,a_{\ell ij\nu_j},0]$ for $\nu_i \ge \nu_j$ or $[a_{\ell ij2},\ldots,a_{\ell ij\nu_i},0,\ldots,0]$ for $\nu_i < \nu_j$. ∎

Corresponding to Corollaries 4.1 and 4.2, respectively, we have:

Corollary 4.3   Let T_0 and T_{RO} be the similarity transformation matrices which transform (A,C) and (A_R,C_R) to their observer forms.   Then

$$S = T_0 \hat{W} T_{RO}^{-1} \qquad\qquad (4.20)$$

where $\hat{W} = T_{0p} \hat{H}^T T_{ROp}^{-1}$; $T_{0p} = T_0^{-1} \mathcal{Q}^{-1}(A,C)$; $T_{ROp}^{-1} = \mathcal{Q}(A_R,C_R)T_{RO}$.   ∎

Corollary 4.4   Let $D_{R\ell}(\lambda)$ be a canonical right divisor of $D_\ell(\lambda)$ defined in Theorem 4.5 and $D_\ell(\lambda) = L(\lambda)D_{R\ell}(\lambda)$.   Then

(1) $L(\lambda) = \psi_\ell(\lambda)\hat{W}\psi_{R\ell}^T(0) + \delta_\ell(\lambda)D_{Lh}(I_m - \psi_{R\ell}(0)\psi_{R\ell}^T(0))$ where $D_{Lh} = D_{\ell h}D_{R\ell h}^{-1}$ is a low triangular matrix with diagonal elements all 1's.

(2) $L(\lambda)$ is a nonsingular λ-matrix.   ∎

The structure theorems of the complete sets of canonical right divisors defined in Definition 4.3 can now be given as follows.

Theorem 4.7   Let S_1 and S_2 be complemented A-invariant subspaces of $\mathcal{X}$(where $S_1 \cap S_2 = \phi$ and $\mathcal{X} = S_1 \oplus S_2$), and (A_{Ri},C_{Ri}),i=1,2, be the embedded observable pairs of (A,C) in S_1,i=1,2.   Then the following diagram commutes.

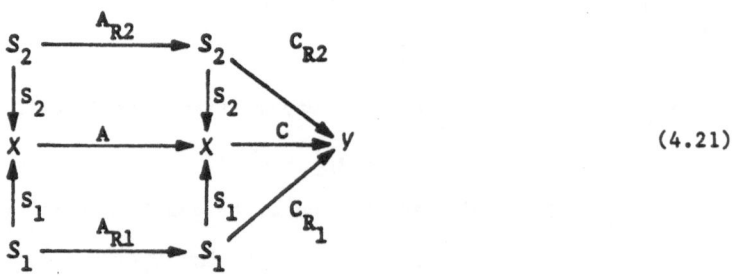

$$(4.21)$$

The set of the left characteristic λ-matrices, $\{D_{R\ell_i}(\lambda),i=1,2\}$, of (A_{Ri},C_{Ri}), i=1,2 is a complete set of canonical right divisors of $D_\ell(\lambda)$.

Proof:

From Theorem 4.5, $D_{R\ell_i}$, i=1,2 are canonical right divisors of $D_\ell(\lambda)$. Since $AS_i = S_i A_{Ri}$, we obtain

$$A[S_1, S_2] = [S_1, S_2] \cdot \text{block diag}[A_{R1}, A_{R2}] .$$

Because $\text{Im}(S_i) = S_i$ and $S_1 \cap S_2 = \phi$, we have $\text{Im}(S_1) \cap \text{Im}(S_2) = \phi$. Also, $S_1 \oplus S_2 = \text{Im}(S_1) \oplus \text{Im}(S_2) = X$, therefore $\text{rank}(S_1) + \text{rank}(S_2) = \dim(X) = n$.

Since $\text{Im}([S_1, S_2]) = \text{Im}(S_1) + \text{Im}(S_2) = \text{Im}(S_1) \oplus \text{Im}(S_2) = X$, or $\text{rank}([S_1, S_2]) = \dim(X) = n$, and $[S_1, S_2]$ is nonsingular, therefore, we obtain $[S_1, S_2]^{-1} A[S_1, S_2] = \text{block diag}[A_{R1}, A_{R2}]$. Thus,

$$\sigma(D_{R\ell_1}(\lambda)) \cup \sigma(D_{R\ell_2}(\lambda)) = \sigma(A_{R1}(\lambda)) \cup \sigma(A_{R2}(\lambda)) = \sigma(\text{block diag}(A_{R1}, A_{R2}))$$

$$= \sigma(A) = \sigma(D_\ell(\lambda)) .$$

Now, we can conclude that $\{D_{R\ell_i}(\lambda), i=1,2\}$ is a complete set of canonical right divisors of $D_\ell(\lambda)$. ∎

The results of Theorem 4.7 can be extended to the following more general case:

Theorem 4.8   Let S_i, i=1,...,k be independent A-invariant subspaces of X such that $X = \bigoplus_{i=1}^{k} S_i$, and (A_{Ri}, C_{Ri}), i=1,...,k be the embedded observable pairs of (A,C) in S_i, i=1,...,k. Also, let $D_{R\ell_i}(\lambda)$, i=1,...,k be the left characteristic λ-matrices of (A_{Ri}, C_{Ri}). Then $\{D_{R\ell_i}(\lambda), i=1,...,k\}$ is a complete set of canonical right divisors of $D_\ell(\lambda)$. ∎

4.3 Structures Theorems for Spectral Factorization of Nonsingular λ-matrices

We now investigate structure theorems for spectral factorization [50-53] of nonsingular λ-matrices.

Let (A,B,C,D) be a minimal realization of $A^{-1}(\lambda)$, where $A(\lambda)\epsilon C^{m\times m}[\lambda]$ is a nonsingular λ-matrix. Let $X\epsilon C^{n}$, $U\epsilon C^{m}$ and $Y\epsilon C^{m}$ be the state, input, and output spaces, respectively. $S\epsilon C^{n_s}$ is an A-invariant subspace of X. (A_L,B_L) and (A_R,C_R) are the induced and embedded pairs in X/S and S, via the epic map V and the monic map S, of (A,B) and (A,C), respectively. Then the following diagram commutes.

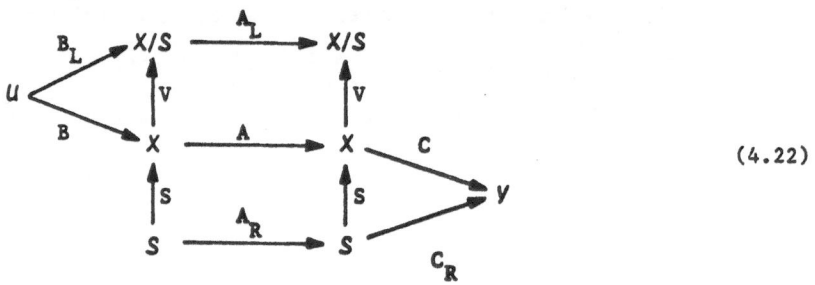

$$(4.22)$$

The sequence $S\to X\to X/S$ is exact at X, i.e. $VS = 0_{(n-n_s)\times n_s}$. Also, (A_L,B_L) and (A_R,C_R) are reachable and observable pairs, respectively.

From the epic map V and the left inverse of the monic map S, we can construct a transformation T to block-triangularize the system map A:

<u>Lemma 4.7</u>   Let $S^+ = (S^*S)^{-1}S^*$ and $V^+ = V^*(VV^*)^{-1}$ be the left and right inverses of S and V, respectively. Define

$$T = \begin{bmatrix} S^+ \\ V \end{bmatrix} \qquad (4.23a)$$

Then

(a) T is nonsingular and $T^{-1} = [S, V^T]$
(b) Let $A_T = TAT^{-1}$, then

$$A_T = \begin{bmatrix} A_R & S^+AV^+ \\ 0 & A_L \end{bmatrix} ; \; A_R \epsilon C^{n_s \times n_s}; \; A_L \epsilon C^{(n-n_s) \times (n-n_s)} \tag{4.23b}$$

(c) $\quad \sigma(A) = \sigma(A_L) \cup \sigma(A_R)$ \hfill (4.23c)

Proof:

(a) Let $\tilde{T} = [S, V^+]$. Then

$$T\tilde{T} = \begin{bmatrix} S^+S & S^+V^+ \\ VS & VV^+ \end{bmatrix} = \begin{bmatrix} I_{n_s} & S^+V^+ \\ VS & I_{n-n_s} \end{bmatrix} = I_n$$

so that $T^{-1} = [S, V^+]$

(b) $\quad A_T = \begin{bmatrix} S^+ \\ V \end{bmatrix} A[S, V^+] = \begin{bmatrix} S^+AS & S^+AV^+ \\ VAS & VAV^+ \end{bmatrix}$

Since $A_L V = VA$ and $SA_R = AS$, we obtain $S^+AS = A_R$; $VAS = VSA_R = 0$; $VAV^+ = A_L$, and therefore, the result in Eq. (4.23b) follows.

(c) The result in Eq. (4.23c) can be proven directly from Eq. (4.23b). ∎

From Theorems 4.1, and 4.4, and Lemma 4.7, we arrive at the following results on the spectral factorization of nonsingular λ-matrices:

Theorem 4.9   Suppose (A,B,C,D) is a minimal realization quadruple of the inverse of a nonsingular λ-matrix, $A(\lambda)$, and the asociated (A_R, C_R) and (A_L, B_L) are defined in Eq. (4.22). Then $A(\lambda)$ can be factored as

$$A(\lambda) = D_L(\lambda) U_A(\lambda) D_R(\lambda) \tag{4.24}$$

where $D_L(\lambda)$ and $D_R(\lambda)$ are the right characteristic λ-matrix of (A_L, B_L) and the left characteristic λ-matrix of (A_R, C_R), respectively. $U_A(\lambda)$ is a unimodular λ-matrix.

Proof:

From Theorems 4.1 and 4.5, we have.

$$A(\lambda) = D_L(\lambda)R(\lambda) = L(\lambda)D_R(\lambda) \qquad (4.25)$$

From Lemma 4.7, there are no shared latent roots between $D_L(\lambda)$ and $D_R(\lambda)$. Thus $D_R(\lambda)$ must be a right divisor of $R(\lambda)$, or $R(\lambda) = U_A(\lambda) \cdot D_R(\lambda)$ and $A(\lambda) = D_L(\lambda)U_A(\lambda)D_R(\lambda)$. Now, we obtain $\det(A(\lambda)) = \det(D_L(\lambda))\det(U_A(\lambda))\det(D_R(\lambda))$. Thus from Lemma 4.7 we observe that $\det(U_A(\lambda))$ is a nonzero constant, and therefore $U_A(\lambda)$ is a unimodular λ-matrix. ∎

When a column/row-reduced non-canonical λ-matrix, $A(\lambda)$, is given and the structures of the column/row-reduced, canonical left/right divisors of $A(\lambda)$ are of interest, we develop the following results.

Theorem 4.10   Let (A,B,C,D) be a minimal realization of the inverse of a column-reduced λ-matrix, $A(\lambda)$, where $A \epsilon C^{n \times n}$, $B \epsilon C^{n \times m}$, $C \epsilon C^{m \times n}$ and $D \epsilon C^{m \times m}$. Let (A_L, B_L) be a reachable pair, where $A_L \epsilon C^{n_v \times n_v}$, $B_L \epsilon C^{n_v \times m}$ and $n \geq n_v > 0$. Also, let $D_{Lr}(\lambda)$ be the right characteristic λ-matrix of (A_L, B_L). Then, $D_{Lr}(\lambda)$ is a canonical left divisor of $A(\lambda)$ iff (A_L, B_L) is an induced reachable pair of (A,B).

Proof:

Since $A(\lambda) = D_r(\lambda)U_r(\lambda)$, from Theorem 4.1, the results of Theorem 4.10 follow. ∎

Corollary 4.5   Let (A,B,C,D) be a minimal realization of the inverse of a column-reduced λ-matrix, $A(\lambda)$, $\hat{T}_c$ be the transformation matrix which transforms (A,B) into a controller form ($\hat{A}_c = \hat{T}_c A \hat{T}_c^{-1}$, $\hat{B}_c = \hat{T}_c B$). Also, let $D_{Lr}(\lambda)$ be the right characteristic λ-matrix of an induced reachable pair (A_L,B_L) of (A,B) via the canonical projection $V\epsilon C^{n_v \times n}$. Then

$$A(\lambda) = D_{Lr}(\lambda)R(\lambda) \qquad (4.26)$$

where

$$D_{Lr}^{-1}(\lambda) = \overline{C}_L(\lambda I_{n_v}-A_L)^{-1}B_L+\overline{D}_L; \quad n_v = \text{rank}(V)$$

$$R(\lambda) = \overline{C}_L V\hat{T}_c^{-1}\psi_c(\lambda)+\overline{D}_L A(\lambda)$$

and $\psi_c(\lambda)$ is defined in Lemma 3.1.

Proof:

From Lemma 3.1 and Eq. (3.2), we obtain

$$\hat{B}_c A(\lambda) = (\lambda I_{n_v}-\hat{A}_c)\psi_c(\lambda)$$

or

$$BA(\lambda) = (\lambda I_n-A)\hat{T}_c^{-1}\psi_c(\lambda) .$$

From Eq. (4.2), we have

$$B_L A(\lambda) = (\lambda I_{n_v}-A_L)V\hat{T}_c^{-1}\psi_c(\lambda) .$$

Thus

$$D_{Lr}^{-1}(\lambda)A(\lambda) = \overline{C}_L V\hat{T}_c^{-1}\psi_c(\lambda)+\overline{D}_L A(\lambda) = R(\lambda) \qquad \blacksquare$$

As usual, the dual results on the spectral factorization of row-reduced λ-

matrices can be written down at once:

Theorem 4.11   Let (A,B,C,D) be a minimal realization of the inverse of a row-reduced λ-matrix, $A(\lambda)$, where $A \varepsilon C^{n \times n}$, $B \varepsilon C^{n \times p}$, $C \varepsilon C^{p \times n}$ and $D \varepsilon C^{p \times p}$. Let (A_R, C_R) be an observable pair, where $A_R \varepsilon C^{n_s \times n_s}$, $C_R \varepsilon C^{p \times n_s}$, $n \geq n_s > 0$. Also, let $D_{R\ell}(\lambda)$ be the left characteristic λ-matrix of (A_R, C_R). Then, $D_{R\ell}(\lambda)$ is a canonical right divisor of $A(\lambda)$ iff (A_R, C_R) is an embedded pair of (A,C).   ■

Corollary 4.6   Let (A,B,C,D) be a minimal realization of the inverse of a row-reduced λ-matrix, $A(\lambda)$ and let $\hat{T}_0$ be the transformation matrix which transforms (A,C) into an observer form $(\hat{A}_0, \hat{C}_0)$, where $\hat{A}_0 = \hat{T}_0^{-1} A \hat{T}_0$, $\hat{C}_0 = C \hat{T}_0$. Also, let $D_{R\ell}(\lambda)$ be the left characteristic λ-matrix of an embedded observable pair (A_R, C_R) of (A,C) via the canonical injection S.   Then,

$$A(\lambda) = L(\lambda) D_{R\ell}(\lambda) \tag{4.27}$$

where

$$D_{R\ell}(\lambda) = C_R (\lambda I_{n_s} - A_R)^{-1} \bar{B}_R + \bar{D}_R; \quad n_s = \text{rank}(S)$$

$$L(\lambda) = \psi_0(\lambda) \hat{T}_0^{-1} S \bar{B}_R + A(\lambda) \bar{D}_R ,$$

and $\psi_0(\lambda)$ is defined in Lemma 3.3.   ■

Theorem 4.11 and Corollary 4.6 can be easily proved by using the same approaches as given in Theorem 4.10 and Corollary 4.5.

4.4 Computational Aspects of Divisors and Spectral Factors for λ-matrices

In this section, we present schemes for computing left/right divisors and spectral factors of nonsingular λ-matrices, and their associated complete sets of irreducible left/right divisors and spectral factors.   The computational schemes are mainly based on block triangularization or block diagonalization [54,55] for the system map of the minimal realization quadruple for the inverse of a nonsingular λ–matrix. At the end of this section, a numerical method for block triangularization and block diagonalization of a square matrix, using the matrix sign algorithm [56-60], is presented.

Let (A,B,C,D) be a minimal realization for the inverse of a nonsingular λ–matrix A(λ). Assume that A can be upper block triangularized as

$$A = \begin{bmatrix} A_1 & A_{12} \\ 0 & A_2 \end{bmatrix} ; \ A_i \in C^{n_i \times n_i}; \ i=1,2; \ A_{12} \in C^{n_1 \times n_2} \tag{4.28a}$$

and B,C can be written as

$$B = [B_1^T, B_2^T]^T; \ B_i \in C^{n_i \times m}; \ i=1,2 \tag{4.28b}$$

$$C = [C_1, C_2]; \ C_i \in C^{m \times n_i}; \ i=1,2 \tag{4.28c}$$

Then, we obtain the following results.

<u>Theorem 4.12</u>   If (A,B,C,D) is a minimal realization of the inverse of a nonsingular λ–matrix A(λ), with A, B and C shown in Eq. (4.28), then

(1) The right characteristic λ–matrix of (A_2, B_2), defined as $D_{r2}(\lambda)$, is a canonical left divisor of A(λ).

(2) The left characteristic λ–matrix of (A_1, C_1), defined as $D_{\ell 1}(\lambda)$, is a canonical right divisor of A(λ).

(3) $A(\lambda) = D_{r2}(\lambda) U_{r\ell}(\lambda) D_{\ell 1}(\lambda)$ where $U_{r\ell}(\lambda)$ is a unimodular λ–matrix.

<u>Proof:</u>

From Theorem 3.3, $A(\lambda) = D_r(\lambda)U_r(\lambda) = U_\ell(\lambda)D_\ell(\lambda)$, where $D_r(\lambda)$ and $D_\ell(\lambda)$ are column- and row-reduced canonical λ-matrices, respectively. Each part of the theorem is established as follows:

(1) Choose $V = [0, I_{n_2}] \epsilon C^{n_2 \times n}$ and apply Theorem 4.1.

(2) Choose $S = [I_{n_1}, 0]^T \epsilon C^{n \times n_1}$ and apply Theorem 4.5.

(3) From Theorem 4.9 and the results in (1) and (2). ∎

Similar results can be stated for a lower block triangularized system map A of a minimal realization for the inverse of a nonsingular λ-matrix as following.

<u>Corollary 4.7</u>   If (A,B,C,D) is a minimal realization for the inverse of a nonsingular λ-matrix, $A(\lambda)$, with A,B and C being partitioned as

$$A = \begin{bmatrix} A_1 & 0 \\ A_{21} & A_2 \end{bmatrix} ; \; A_i \epsilon C^{n_i \times n_i}; \; i=1,2; \; A_{21} \epsilon C^{n_2 \times n_1} \tag{4.29a}$$

$$B = [B_1^T, B_2^T]^T; \; B_i \epsilon C^{n_i \times m}; \; i=1,2 \tag{4.29b}$$

$$C = [C_1, C_2]; \; C_i \epsilon C^{m \times n_i}; \; i=1,2 \tag{4.29c}$$

then

(1) The right characteristic λ-matrix of (A_1, B_1), defined as $D_{r1}(\lambda)$, is a canonical left divisor of $A(\lambda)$.

(2) The left characteristic λ-matrix of (A_2, C_2), defined as $D_{\ell2}(\lambda)$, is a canonical right divisor of $A(\lambda)$.

(3) $A(\lambda) = D_{r1}(\lambda)U_{r\ell}(\lambda)D_{\ell2}(\lambda)$ where $U_{r\ell}(\lambda)$ is a unimodular λ-matrix. ∎

To find a complete set of canonical left or right divisors, we need to block-diagonalize the system map of A of the minimal realization quadruple for the inverse of a nonsingular λ-matrix.

__Theorem 4.13__   If (A,B,C,D) is a minimal realization of $A^{-1}(\lambda)$ and A is block-diagonalized as

$$A = \text{block diag}[A_i, i=1,\ldots,k]; \quad A_i \epsilon C^{n_i \times n_i}; \quad i=1,\ldots,k \qquad (4.30a)$$

and

$$B = [B_1^T, B_2^T, \ldots, B_k^T]; \quad B_i \epsilon C^{n_i \times m}; \quad i=1,\ldots,k \qquad (4.30b)$$

$$C = [C_1, C_2, \ldots, C_k]; \quad C_i \epsilon C^{m \times n_i}; \quad i=1,\ldots,k \qquad (4.30c)$$

then

(1)   The right characteristic λ-matrices of (A_i, B_i), $D_{ri}(\lambda)$, for $1 \le i \le k$, constitute a complete set of canonical left divisors of $A(\lambda)$.

(2)   The left characteristic λ-matrices of (A_i, C_i), $D_{\ell i}(\lambda)$, for $1 \le i \le k$, constitute a complete set of canonical right divisors of $A(\lambda)$.

__Proof:__

(1)   Choose $V_i = \text{block diag}[0_{n_1}, \ldots, 0_{n_i-1}, I_{n_i}, 0_{n_i+1}, \ldots, 0_{n_k}]$. From Theorem 4.12, we obtain $D_{ri}(\lambda)$ as a canonical left divisor of $A(\lambda)$ for $i=1,\ldots,k$. Also, using Lemma 2.1 yields $\sigma(D_{ri}(\lambda)) = \sigma(A_i)$, and hence $\bigcup_{i=1}^{k} \sigma(D_{ri}(\lambda)) = \bigcup_{i=1}^{k} \sigma(A) = \sigma(A(\lambda))$. Thus, $\{D_{ri}(\lambda), i=1,\ldots,k\}$ is a complete set of canonical left divisors of $A(\lambda)$.

(2)   Similar to the proof in (1).   ∎

The computational scheme for the irreducible complete set of canonical left/right divisors for a nonsingular λ-matrix $A(\lambda)$ can be formulated by using the Jordan canonical form minimal realization of $A^{-1}(\lambda)$.

__Theorem 4.14__   If (A,B,C,D) is a minimal realization of $A^{-1}(\lambda)$ and A is in Jordan canonical form as

$$A = \text{block diag}[A_{Ji}, i=1,\ldots,k]$$

where $A_{Ji} \epsilon C^{n_i \times n_i}$ is a Jordan block associated with an eigenvalue λ_i and a full Jordan chain as

$$A_{Ji} = \begin{bmatrix} \lambda_i & 1 & \cdot & \cdot \\ \cdot & \lambda_i & 1 & \cdot \\ \cdot & \cdot & \cdot & \cdot \\ \cdot & \cdot & \cdot & \lambda_i \end{bmatrix} \qquad (4.31a)$$

and B and C are represented as

$$B = [B_{J1}^T,\ldots,B_{Jk}^T]^T; \ B_{Ji} \epsilon C^{n_i \times m}; \ i=1,\ldots,k \qquad (4.31b)$$

$$C = [C_{J1},\ldots,C_{Jk}]; \ C_{Ji} \epsilon C^{m \times n_i}; \ i=1,\ldots,k \qquad (4.31c)$$

then,

(1) The right characteristic λ-matrices of (A_{Ji},B_{Ji}), $D_{ri}(\lambda)$ for $1 \le i \le k$, constitute an irreducible complete set of canonical left divisors of $A(\lambda)$. Also, $\{D_{ri}(\lambda), i=1,\ldots,k\}$ is unique.

(2) The left characteristic λ-matrices of (A_{Ji},C_{Ji}), $D_{\ell i}(\lambda)$ for $1 \le i \le k$, constitute an irreducible complete set of canonical right divisors of $A(\lambda)$. Also, $\{D_{\ell i}(\lambda), i=1,\ldots,k\}$ is unique.

<u>Proof:</u>

(1) From Theorem 4.13, $\{D_{ri}(\lambda),i=1,\ldots,k\}$ is a complete set of canonical left divisors of $A(\lambda)$. Since each A_{Ji} contains a Jordan block with a full Jordan chain, A_{Ji} cannot be further decomposed into more than one block. This fact shows that $\{D_{ri}(\lambda),i=1,\ldots,k\}$ is an irreducible complete set of canonical left divisors of $A(\lambda)$. Moreover, the Jordan block decomposition of a square matrix is unique, and therefore the irreducible complete set of left divisors, $\{D_{ri}(\lambda),i=1,\ldots,k\}$, is unique.

(2) Similar to (1). ∎

From Theorems 4.12 and 4.13, we observe that the canonical left/right divisors and complete sets of canonical left/right divisors of nonsingular λ-matrices can be determined from the block triangularization/diagonalization of the system map A of the minimal realization quadruple (A,B,C,D) for the inverse of the λ-matrices. The block triangularization and diagonalization of a square matrix can be reformulated as an algebraic Riccati equation problem [54] and several numerical algorithms are available for solving such Riccati equations [55-56]. Here, we shall present a new algorithm for block diagonalization and block triangularization of a square matrix using the matrix sign algorithm [60]. Let us define the matrix sign function first.

<u>Definition 4.4</u>   The <u>matrix sign function</u> of $A\epsilon C^{n\times n}$ with $\sigma(A)\subset C^+\cup C^-$, where C^+ and C^- are the open right and left plane of C, respectively, is defined by [56]

$$\text{Sign}(A) = 2\,\text{Sign}^+(A)-I_n = I_n-2\,\text{Sign}^-(A) \tag{4.32a}$$

where

$$\text{Sign}^+(A) = \frac{1}{2\pi i}\oint_{\Gamma_+}(\lambda I_n-A)^{-1}d\lambda \tag{4.32b}$$

and

$$\text{Sign}^-(A) = \frac{1}{2\pi i} \oint_{\Gamma_-} (\lambda I_n - A)^{-1} d\lambda \tag{4.32c}$$

$\Gamma_+(\Gamma_-)$ is a simple closed contour in $C^+(C^-)$, and encloses $\sigma(A) \cap C^+$ ($\sigma(A) \cap C^-$). $\quad\blacksquare$

From Definition 4.4, we can formulate Sign(A) using the modal matrix M of A. Let

$$J = M^{-1}AM = \text{block diag}[J_+, J_-] \tag{4.33a}$$

where $J_+ \in C^{n_1 \times n_1}$ and $J_- \in C^{n_2 \times n_2}$ ($n = n_1 + n_2$) are the collections of Jordan blocks with $\sigma(J_+) \subset C^+$ and $\sigma(J_-) \subset C^-$, respectively. Then,

$$\text{Sign}(A) = M[\text{Sign}(J_+) \oplus \text{Sign}(J_-)]M^{-1} = M[I_{n_1} \oplus (-I_{n_2})]M^{-1} \tag{4.33b}$$

Assume that M is partitioned as

$$M = [M_1, M_2], \quad M_1 \in C^{n \times n_1}, \quad M_2 \in C^{n \times n_2} \tag{4.34}$$

where M_1 and M_2 contain the eigenvectors associated with the eigenvalues of J_+ and J_-, respectively. Let W be the inverse of M and be partitioned as

$$M^{-1} \triangleq W = \begin{bmatrix} W_1 \\ W_2 \end{bmatrix}, \quad W_1 \in C^{n_1 \times n}, \quad W_2 \in C^{n_2 \times n} \tag{4.35}$$

Then, from Eq. (4.32a), we obtain

$$\text{Sign}^+(A) = \frac{1}{2}[\text{Sign}(A) + I_n] = M[I_{n_1} \oplus 0_{n_2}]M^{-1} = M_1 W_1 \tag{4.36a}$$

Obviously, $\mathrm{rank}[\mathrm{Sign}^+(A)] = n_1$. Let

$$\mathrm{Sign}^+(A) = [s_1, s_2, \ldots, s_n] \in C^{n \times n}, \ s_i \in C^{n \times 1} \tag{4.36b}$$

and

$$S \overset{\Delta}{=} \mathrm{Ind}[\mathrm{Sign}^+(A)] \in C^{n \times n_1} \tag{4.37}$$

where S is a monic map which contains n_1 **independent** column vectors of $\mathrm{Sign}^+(A)$. These independent vectors are selected from the $n(>n_1)$ column vectors of $\mathrm{Sign}^+(A)$ in Eq. (4.36b).

Thus, we obtain the results as follows.

Lemma 4.8 There exists a nonsingular matrix $\xi \in C^{n_1 \times n_1}$ such that $S = M_1 \xi$ and $S^+ = \xi^{-1} M_1^+$, where M_1 is defined in Eq. (4.34); $S^+ \overset{\Delta}{=} (S^*S)^{-1}S^*$ and $M_1^+ \overset{\Delta}{=} (M_1^*M_1)^{-1}M_1^*$ are the left inverses of M_1 and S, respectively.

Proof:

Assume that S contains n_1 columns of $\mathrm{Sign}^+(A)$ with column indices as k_i for $i=1,2,\ldots,n_1$. Then from the definition of S we obtain

$$S = \mathrm{Sign}^+(A)U_s$$

where $U_s \overset{\Delta}{=} [e_n^{k_1}, e_n^{k_2}, \ldots, e_n^{k_n}]$.

Since $\mathrm{Sign}^+(A) = M_1 W_1$, we have $S = M_1 W_1 U_s \overset{\Delta}{=} M_1 \xi$, where $\xi \overset{\Delta}{=} W_1 U_s$. By Sylvester's inequality [61], we obtain

$$\mathrm{rank}(W_1) + \mathrm{rank}(U_s) - n_1 \le \mathrm{rank}(\xi) \le \min(\mathrm{rank}(W_1), \mathrm{rank}(U_s))$$

and since $\mathrm{rank}(W_1) = \mathrm{rank}(U_s) = n_1$, we conclude that

$$\mathrm{rank}(\xi) = n_1 \tag{4.38}$$

From the definition of the left inverse of S^+, we have

$$S^+ = (S^*S)^{-1}S^* = [(M_1\xi)^*(M_1\xi)]^{-1}(M_1\xi^*) = \xi^{-1}M_1^+ \qquad \blacksquare$$

Two important properties of S are given by:

<u>Theorem 4.15</u>   Define $S \overset{\Delta}{=} Im(S)$ and let X be the state space and A: $X \rightarrow X$.   Then,

(i) S is an A-invariant subspace of X.

(ii) S is the canonical injection map S: $S \rightarrow X$.

<u>Proof</u>:

(i)   From Lemma 4.8, we obtain

$$AS = AM_1\xi = M_1 J_+\xi = M_1\xi(\xi^{-1}J_+\xi) = SA_R \qquad (4.39a)$$

where
$$A_R \overset{\Delta}{=} \xi^{-1}J_+\xi \qquad (4.39b)$$

Thus, $S = Im(S)$ is an A-invariant subspace of X.

(ii) Since S is monic and $S = Im(S)$, S is the canonical injection map S: $S \rightarrow X$. $\blacksquare$

Obviously, from Theorems 4.5 and 4.15, S defined in Eq. (4.37) can be used as a canonical injection map for finding the canonical right divisor of nonsingular λ-matrices.   Since S is composed of the independent column vectors of $Sign^+(A)$, the orthogonalized projection algorithm in Section 2.3 can be used to find S from $\bar{S} \overset{\Delta}{=} Sign^{-1}(A)$:

__Algorithm 4.1__

Given: $\bar{S} = [S_1, S_2, \ldots, S_n], S_i \epsilon C^{n \times 1}$, for $1 \leq i \leq n$

Find: $S = [S_{k1}, S_{k2}, \ldots, S_{kn_1}]$, such that $S_{ki} \epsilon \{S_i, 1 \leq i \leq n\}$ for $1 \leq k_i \leq n_1$ are independent.

Algorithm:

    {Initialization}

    j:=0;{set independent vector index starting at zero}

    $P:=I_n$;{set orthogonalized projection to identity matrix}

    {Processing}

    For i=1 to n Do

      Begin

        $d:=P*S_i$;

        If $S_i^* *d \neq 0$ Then

          Begin

            j:=j+1;{increment index j}

            $S_{kj}:=S_i$;{v_i is an linearly independent column vector}

            $P:=P-d*d^*/S_i^**d$;{Update P}

          End{If}

      End;{For loop}

    $n_1:=j$;{Total number of linearly independent column vectors which is the rank of $\bar{S}$}

Note: The * between two variables in the above algorithm is the matrix product notation; and the superscript * designates conjugate transpose.   ■

The matrix S defined in Eq. (4.37) can be used for block diagonalization of a system map as follows.

<u>Theorem 4.16</u>   Let $A \epsilon C^{n \times n}$ with $Re(\lambda_i) \neq 0$ for $i=1,2,\ldots n$, where $\{\lambda_i, i=1,2,\ldots,n\}$ is $\sigma(A)$. Define

$$S_1 = \text{Ind}[\text{Sign}^+(A)] \epsilon C^{n \times n_1} \tag{4.40a}$$

$$S_2 = \text{Ind}[\text{Sign}^-(A)] \epsilon C^{n \times n_2} \tag{4.40b}$$

where $\text{Sign}^+(A)$ and $\text{Sign}^-(A)$ are defined in Eq. (4.42) and $n_1 + n_2 = n$.

Assume that $n_1 n_2 \neq 0$, then

$$A_D = M_s^{-1} A M_s = \text{block diag}[A_{R1}, A_{R2}] \tag{4.41a}$$

where M_s is a block modal matrix and expressed as

$$M_s \triangleq [S_1, S_2] \epsilon C^{n \times n} \tag{4.41b}$$

and $A_{R1} \epsilon C^{n_1 \times n_1}$ and $A_{R2} \epsilon C^{n_2 \times n_2}$ are defined as

$$A_{R1} = S_1^+ A S_1 \epsilon C^{n_1 \times n_1} \tag{4.41c}$$

$$A_{R2} = S_2^+ A S_2 \epsilon C^{n_2 \times n_2} \tag{4.41d}$$

S_1^+ and S_2^+ are the left inverses of S_1 and S_2, or $S_1^+ = (S_1^* S_1)^{-1} S_1^*$ and $S_2^+ = (S_2^* S_2)^{-1} S_2^*$, respectively, where $*$ designates conjugate transpose.

<u>Proof:</u>

Theorem 4.16 can be proved using Theorem 4.15 and Eq. (4.39).   ∎

If more than two blocks are needed for the block diagonalization of A, we can construct the block modal matrix using generalized sign matrix functions.

Definition 4.5   Let $A \epsilon C^{n \times n}$ and $Re(\sigma(A)) \cap \{r_1, r_2\} = \phi$, where $r_1 < r_2$ and $r_1, r_2 \epsilon R$. The generalized matrix sign function of A with respect to the open interval (r_1, r_2) on the real axis is defined as

$$Sign_{(r_1, r_2)} = 2\ Sign^+_{(r_1, r_2)}(A) - I_n$$

$$= I_n - 2\ Sign^-_{(r_1, r_2)}(A) \tag{4.42a}$$

where

$$Sign^+_{(r_1, r_2)}(A) \overset{\Delta}{=} \frac{1}{2}[Sign_{(r_1)}(A) - Sign_{(r_2)}(A)] \tag{4.42b}$$

$$Sign^-_{(r_1, r_2)}(A) \overset{\Delta}{=} I_n - Sign^+_{(r_1, r_2)}(A) \tag{4.42c}$$

and

$$Sign_{(r_i)}(A) \overset{\Delta}{=} Sign(A - r_i I_n), \text{ for } i=1,2 \ . \tag{4.42d}$$

□

Theorem 4.17   Let $A \epsilon C^{n \times n}$ and $Re(\sigma(A)) \cap \{r_i, i=1, \ldots, k\} = \phi$, where $r_1 < r_2 < \ldots < r_k$ and $r_i \epsilon R$ for $1 \leq i \leq k$. Define

$$S_i = Ind(Sign^+_{(r_{i-1}, r_i)}(A)) \epsilon C^{n \times n_i}, \ 1 \leq i \leq k+1 \tag{4.43}$$

where $r_0 \overset{\Delta}{=} -\infty$ and $r_{k+1} \overset{\Delta}{=} \infty$. Assume that $n_i \neq 0$ for $1 \leq i \leq k+1$. Then

$$M_s^{-1} A M_s = \text{Block diag}[A_{R1}, A_{R2}, \ldots, A_{R(k+1)}] \tag{4.44a}$$

where M_s is a block modal matrix

$$M_s = [S_1, S_2, \ldots, S_{k+1}] \tag{4.44b}$$

and A_{Ri}, $1 \leq i \leq k+1$, are defined as

$$A_{Ri} = S_i^+ A S_i \tag{4.44c}$$

where S_i^+ is the left inverse of S_i for $1 \leq i \leq k+1$.

Proof:

Directly from Theorem 4.15 and Definition 4.5. From Definition 4.5 and Theorem 4.17, we have the following:

Corollary 4.8   Let A_{Ri} be defined as in Eq. (4.44c). Then $r_{i-1} < \text{Re}(\lambda_{ij}) < r_i$, for all eigenvalues λ_{ij} of A_{Ri}, $1 \leq j \leq n_i$.   ∎

Theorems 4.13 and 4.17, together with Corollary 4.9, allow us to decompose a nonsingular λ-matrix into a complete set of canonical right divisors such that each divisor has latent roots clustering inside a certain vertical strip of the complex plane C.

To obtain the canonical projection map from matrix sign function, we define

$$V = \{\text{Ind}[(\text{Sign}^+(A))^T]\}^T \tag{4.45}$$

which is epic. Note that V contains n_1 independent row vectors of $\text{Sign}^+(A)$ and these independent vectors are selected from the $n(>n_1)$ row vectors of $\text{Sign}^+(A)$ in Eq. (4.36c). Parallel to Lemma 4.8, we have:

Lemma 4.9   There exists a nonsingular matrix $\eta \epsilon C^{n_1 \times n_1}$ such that $V = \eta W_1$ and V^+

$= W_1^+ \eta^{-1}$, where W_1 is defined in Eq. (4.35); $V^+ \overset{\Delta}{=} V^*(VV^*)^{-1}$ and $W_1^+ \overset{\Delta}{=} W_1^*(W_1 W_1^*)^{-1}$ are the right inverse of V and W_1, respectively.

<u>Proof</u>:

From the definition of V we can describe V as

$$V = U_V \text{Sign}^+(A) \tag{4.46}$$

where $U_V^T \overset{\Delta}{=} [e_n^{k_1}, e_n^{k_2}, \ldots, e_n^{k_{n_1}}]$.

Since from Eq. (4.36b) we have $\text{Sign}^+(A) = M_1 W_1$, Eq. (4.46) becomes $V = U_V M_1 W_1 = \eta W_1$, where $\eta \overset{\Delta}{=} U_V M_1$. By applying Sylvester's inequality [61], we have $\text{rank}(\eta) = n_1$. From the definition of right inverse, we have

$$V^+ = V^*(VV^*)^{-1} = (\eta W_1)^*(\eta W_1 W_1^* \eta^*)^{-1} = W_1^+ \eta^{-1} \tag{4.47}$$

∎

Algorithm 4.1 can be utilized to find V from $\text{Sign}^+(A)$ by replacing $[\text{Sign}^+(A)]^T$ for $\bar{S}$ and V^T for S. Corresponding to Theorem 4.15, the basic properties of V are given by:

<u>Theorem 4.18</u>   Define $S \overset{\Delta}{=} \text{Ker}(V)$; let X be the state space and $A: X \to X$. Then
(i) S is an A-invariant subspace of X.
(ii) V is the canonical projection map $V: X \to X/S$.   ∎

Obviously, from Theorems 4.1 and 4.18, V defined in Eq. (4.45) can be used as the canonical projection map for finding the canonical right divisors of nonsingular λ-matrices.

Paralleling Theorem 4.17, we can construct the block modal matrix from canonical projection maps.

<u>Theorem 4.19</u>   Let $A \in C^{n \times n}$ and $\text{Re}(\sigma(A)) \cap \{r_i, i=1,\ldots,k\} = \phi$, where $r_1 < r_2 < \ldots < r_k$

and $r_i \in R$ for $1 \le i \le k$. Define

$$V_i = \{Ind((Sign^+_{(r_{i-1},r_i)}(A))^T)\}^T \epsilon C^{n_i \times n}, \quad 1 \le i \le k+1 \qquad (4.48)$$

where $r_0 = -\infty$ and $r_{k+1} = \infty$. Assume that $n_i \ne 0$ for $1 \le i \le k+1$. Then

$$M_V A M_V^{-1} = Block\ diag[A_{L1}, A_{L2}, \ldots, A_{L(k+1)}] \qquad (4.49a)$$

where M_V^{-1} is a block modal matrix

$$M_V = [v_1^T, v_2^T, \ldots, v_{k+1}^T]^T \qquad (4.49b)$$

A_{Li}, $1 \le i \le k+1$, are defined as following

$$A_{Li} = V_i A V_i^+ \qquad (4.49c)$$

V_i^+ is the right inverse of V_i. Also, the real parts of all the eigenvalues of A_{Li} are in (r_{i-1}, r_i).

Proof:

Directly from Theorem 4.18 and Definition 4.5. ∎

Since matrix sign functions can be used to compute the canonical injection and canonical projection maps, the block triangularization transformation can be easily obtained from matrix sign functions of A by Lemma 4.7.

Theorem 4.20   Let $A \epsilon C^{n \times n}$ and $Re(\sigma(A)) \cap \{r_1, r_2\} = \phi$, where $r_1 < r_2$ and $r_i \epsilon R$ for $i=1,2$. Define

$$S_{(r_1,r_2)} = \text{Ind}(\text{Sign}^+_{(r_1,r_2)}(A)) \in C^{n \times n_1} \tag{4.50a}$$

and

$$\bar{V}_{(r_1,r_2)} = [\text{Ind}((\text{Sign}^-_{(r_1,r_2)}(A))^T)]^T \in C^{n \times n_2} \tag{4.50b}$$

Assume that n_1 and n_2 are positive. Let

$$T \triangleq \begin{bmatrix} S^+_{(r_1,r_2)} \\ \bar{V}_{(r_1,r_2)} \end{bmatrix} \tag{4.51}$$

where $S^+_{(r_1,r_2)}$ is the left inverse of $S_{(r_1,r_2)}$. Then

$$A_T \triangleq TAT^{-1} = \begin{bmatrix} A_R & A_{RL} \\ 0_{n_2 \times n_1} & A_L \end{bmatrix} \tag{4.52a}$$

where

$$A_R \triangleq S^+_{(r_1,r_2)} A S_{(r_1,r_2)} \tag{4.52b}$$

$$A_L \triangleq \bar{V}_{(r_1,r_2)} A \bar{V}^+_{(r_1,r_2)} \tag{4.52c}$$

$$A_{RL} \triangleq S^+_{(r_1,r_2)} A \bar{V}^+_{(r_1,r_2)} \tag{4.52d}$$

and $\bar{V}^+_{(r_1,r_2)}$ is the right inverse of $\bar{V}_{(r_1,r_2)}$. Furthermore, the real parts of the eigenvalues of A_R are inside (r_1,r_2) and those of A_L are outside (r_1,r_2).

Proof:

From Theorem 4.15, $S_{(r_1,r_2)}$ is the canonical injection map, $S_{(r_1,r_2)}$: $S \rightarrow X$, where $S \triangleq \text{Im}(S_{(r_1,r_2)})$ and X is the state space such that A: $X \rightarrow X$. From the definition of $\bar{V}_{(r_1,r_2)}$ and Lemma 4.9, we have

$$\bar{V}_{(r_1,r_2)} = U_{\bar{V}}(I_n - Sign^+_{(r_1,r_2)}(A))$$

where

$$U_{\bar{V}} = [e_n^{\bar{k}_1}, e_n^{\bar{k}_2}, \ldots, e_n^{\bar{k}_{n_2}}]^T$$

Also from Lemma 4.8, we have

$$S_{(r_1,r_2)} = Sign^-_{(r_1,r_2)}(A)U_s$$

where

$$U_s = [e_n^{k_1}, e_n^{k_2}, \ldots, e_n^{k_{n_1}}]$$

From Eq. (4.36a), it is easy to show that $Sign^-_{(r_1,r_2)}(A)$ is idempotent. Thus, we obtain

$$\bar{V}_{(r_1,r_2)}S_{(r_1,r_2)} = U_{\bar{V}}(I_n - Sign^-_{(r_1,r_2)}(A))Sign^-_{(r_1,r_2)}(A)U_s = 0$$

since $\bar{V}_{(r_1,r_2)}$ and $S_{(r_1,r_2)}$ are of full rank. This implies that $Ker(\bar{V}_{(r_1,r_2)} = Im(S_{(r_1,r_2)} = S$ or $\bar{V}_{(r_1,r_2)}$ is a canonical projection map. By Lemma 4.7, the rest of the results of Theorem 4.20 follow. ∎

To compute $Sign(A)$ or $Sign^+(A)$, Roberts [56] proposed a recursive algorithm, which is referred to as the matrix sign algorithm [56] and can be described as follows:

$$Sign^{(i+1)}(A) = \frac{1}{2}[Sign^{(i)}(A) + (Sign^{(i)}(A))^{-1}]; \quad Sign^{(0)}(A) = A \text{ for } i \geq 0 \quad (4.53a)$$

and

$$Sign(A) = \lim_{i \to \infty} Sign^{(i)}(A) \quad (4.53b)$$

The algorithm in Eq. (4.53a) provides quadratic convergence at the neighborhood of Sign(A). Recently, a new algorithm to compute the matrix sign functions has been devised [60]. This new algorithm for Sign(A) is as follows:

__Theorem 4.21__   The matrix sign function of A can be represented by a matrix continued fraction:

$$\text{Sign}(A) = A[I_n+(A^2-I_n)[2I_n+(A^2-I_n)[2I_n+(A^2-I_n)[...]^{-1}]^{-1}]^{-1}]^{-1} \qquad (4.54a)$$

where $A \varepsilon C^{n \times n}$ and $\text{Re}[\sigma(A)] \neq 0$.

The approximate matrix sign function becomes

$$\text{Sign}(A) \simeq [(I_n+A)^j-(I_n-A)^j][(I_n+A)^j+(I_n-A)^j]^{-1} \triangleq \text{Sign}_{(j)}(A) \qquad (4.54b)$$

where j is a finite integer.

The recursive algorithm for computing the matrix sign function of A with $\text{Re}[\sigma(A)] \neq 0$ is

$$\text{Sign}_{(n_{k+1})}(A) = \text{Sign}_{(f_k)}[\text{Sign}_{(n_k)}(A)] \qquad (4.55)$$

where $\text{Sign}_{(1)}(A) \triangleq A$; $n_{k+1} = f_k \cdot n_k$ and $f_k > 1$ for $k \geq 0$; $n_0 = 1$.   ■

The degree of the convergence rate of the algorithm in Eq. (4.55) at the neighborhood of Sign(A) is f_k if f_k is constant for $k \geq 0$. The reader is referred to [60] for further details of the new matrix sign algorithm in Eq. (4.55).

Theorems 4.15-4.21 present a set of computational algorithms for block diagonalization and block triangularization of a square matrix A such that each block matrix along the diagonal has eigenvalues clustering inside a certain vertical strip of the complex plane C.   These algorithms are suitable for finding the complete sets of left/right divisors and spectral factors of nonsingular λ–matrices occurred in the continuous-time systems (see Chapter VI). For discrete-time system problems we often need to block-diagonalize or block-triangularize a system map A such that each block matrix along the diagonal has eigenvalues clustering inside a certain annulus centered at origin of the complex plane C. Therefore, to use Theorems 4.15-4.21 for discrete-time systems, the argument matrix A in the matrix sign functions, Sign(A), Sign^+(A) and Sign^-(A), should be replaced by $(A-I_n)(A+I_n)^{-1}$; $\text{Sign}_{(r_1,r_2)}$(A), $\text{Sign}^+_{(r_1,r_2)}$(A) and $\text{Sign}^-_{(r_1,r_2)}$(A) in Definition 4.5 should be redefined accordingly:

$$\text{Sign}_{(r_1,r_2)}(A) = 2 \text{ Sign}^+_{(r_1,r_2)}(A) - I_n$$

$$= I_n - 2 \text{ Sign}^-_{(r_1,r_2)}(A) \tag{4.56a}$$

where

$$\text{Sign}^+_{(r_1,r_2)}(A) \triangleq \frac{1}{2}[\text{Sign}_{(r_1)}(A) - \text{Sign}_{(r_2)}(A)] \tag{4.56b}$$

$$\text{Sign}^-_{(r_1,r_2)}(A) \triangleq I_n - \text{Sign}^+_{(r_1,r_2)}(A) \tag{4.56c}$$

$$\text{Sign}_{(r_i)}(A) \triangleq \text{Sign}((A-I_n)(A+I_n)^{-1} - r_i I_n), \text{ for } i=1,2 \tag{4.56d}$$

With the definitions in Eq. (4.56), $\text{Sign}_{(r_1,r_2)}$(A) is called the generalized matrix sign function of A with respect to the open annulus centered at origin of C and bounded with radii $(1+r_1)/(1-r_1)$ and $(1+r_2)/(1-r_2)$.

4.5 Illustrative Examples

Two examples are illustrated in this section. The first example demonstrates the computations of the complete sets of canonical left/right divisors and the spectral factors for a nonsingular λ–matrix via the Jordan form decompositions. The second example illustrates the computations of the complete sets of canonical left/right divisors and spectral factors for a nonsingular λ–matrix via the matrix sign algorithms.

Example 4.1

Consider a reachable pair (A,B) of a continuous-time system:

$$A = \begin{bmatrix} -1 & -2 & 4 & -9 & 21 \\ 0 & -2 & 2 & -5 & 4 \\ 0 & 2 & -5 & 10 & -23 \\ 0 & 1 & -2 & 4 & -4 \\ 0 & 0 & 0 & 0 & 2 \end{bmatrix}$$

and

$$B = \begin{bmatrix} 1 & 3 & 2 \\ 0 & -2 & -2 \\ 0 & 1 & 1 \\ 2 & 5 & 3 \\ 1 & 2 & 1 \end{bmatrix}$$

The right characteristic λ-matrix is found to be

$$D_r(\lambda) = \begin{bmatrix} \lambda^2 - \lambda - 2 & -18\lambda - 18 & 1 \\ 0 & \lambda^3 + 3\lambda^2 + 3\lambda + 1 & -1 \\ 0 & 0 & 1 \end{bmatrix}$$

which is a column-reduced canonical λ-matrix. The Jordan form minimal realization quadruple of $D_r^{-1}(\lambda)$ can be found as

$$A_J = \begin{bmatrix} -1 & 1 & 0 & 0 & 0 \\ 0 & -1 & 1 & 0 & 0 \\ 0 & 0 & -1 & 0 & 0 \\ 0 & 0 & 0 & -1 & 0 \\ 0 & 0 & 0 & 0 & 2 \end{bmatrix}$$

$$B_J = \begin{bmatrix} 1 & 2 & 1 \\ 0 & 1 & 1 \\ 0 & 1 & 1 \\ 1 & 1 & 0 \\ 1 & 2 & 1 \end{bmatrix}$$

$$C_J = \begin{bmatrix} -6 & 4 & 5/3 & 17/3 & 1/3 \\ 1 & -1 & 0 & -1 & 0 \\ 0 & 0 & 0 & 0 & 0 \end{bmatrix}$$

$$D_J = \begin{bmatrix} 0 & 0 & 0 \\ 0 & 0 & 0 \\ 0 & 0 & 1 \end{bmatrix}$$

Let A_J and B_J be partitioned as

$$A_J = \begin{bmatrix} A_1 & 0 \\ 0 & A_2 \end{bmatrix} \qquad B_J = \begin{bmatrix} B_1 \\ B_2 \end{bmatrix}$$

where

$$A_1 = \begin{bmatrix} -1 & 1 & 0 \\ 0 & -1 & 1 \\ 0 & 0 & -1 \end{bmatrix} \qquad A_2 = \begin{bmatrix} -1 & 0 \\ 0 & 2 \end{bmatrix}$$

$$B_1 = \begin{bmatrix} 1 & 2 & 1 \\ 0 & 1 & 1 \\ 0 & 1 & 1 \end{bmatrix} \qquad B_2 = \begin{bmatrix} 1 & 1 & 0 \\ 1 & 2 & 1 \end{bmatrix}$$

The right characteristic λ-matrices of (A_1, B_1) and (A_2, B_2) are

$$D_{r1}(\lambda) = \begin{bmatrix} \lambda+1 & -1 & 1 \\ 0 & \lambda^2+2\lambda+1 & -1 \\ 0 & 0 & 1 \end{bmatrix}$$

and

$$D_{r2}(\lambda) = \begin{bmatrix} \lambda+4 & 6 & 1 \\ -3 & \lambda-5 & -1 \\ 0 & 0 & 1 \end{bmatrix}$$

respectively. From Theorem 4.13, $\{D_{r1}(\lambda), D_{r2}(\lambda)\}$ is a complete set of canonical left divisors of $D_r(\lambda)$. Similarly, let C_J be partitioned as

$$C_J = [C_1, C_2]$$

where

$$C_1 = \begin{bmatrix} -6 & 4 & 5/3 \\ 1 & -1 & 0 \\ 0 & 0 & 0 \end{bmatrix} \qquad C_2 = \begin{bmatrix} 17/3 & 1/3 \\ -1 & 0 \\ 0 & 0 \end{bmatrix}$$

The left characteristic λ-matrices of (A_1, C_1) and (A_2, C_2) are

$$D_{\ell1}(\lambda) = \begin{bmatrix} \lambda^2-4\lambda+13 & 108 & 0 \\ 25/3 & \lambda+7 & 0 \\ 0 & 0 & 1 \end{bmatrix} .$$

and

$$D_{\ell 2}(\lambda) = \begin{bmatrix} \lambda-2 & -17 & 0 \\ 0 & \lambda+1 & 0 \\ 0 & 0 & 1 \end{bmatrix}$$

respectively. From Theorem 4.13, $\{D_{\ell 1}(\lambda), D_{\ell 2}(\lambda)\}$ is a complete set of canonical right divisors of $D_r(\lambda)$.

Let A_J, B_J, C_J be partitioned as

$$A_J = \begin{bmatrix} \bar{A}_1 & \bar{A}_{12} \\ 0 & \bar{A}_2 \end{bmatrix} \quad , \quad B_J = \begin{bmatrix} \bar{B}_1 \\ \bar{B}_2 \end{bmatrix} , \quad C_J = [\bar{C}_1, \bar{C}_2]$$

where

$$\bar{A}_1 = \begin{bmatrix} -1 & 0 \\ 0 & -1 \end{bmatrix} \quad \bar{A}_2 = \begin{bmatrix} -1 & 0 & 0 \\ 0 & -1 & 0 \\ 0 & 0 & 2 \end{bmatrix} \quad \bar{A}_{12} = \begin{bmatrix} 0 & 0 & 0 \\ 1 & 0 & 0 \end{bmatrix}$$

$$\bar{B}_1 = \begin{bmatrix} 1 & 2 & 1 \\ 0 & 1 & 1 \end{bmatrix} \quad \bar{B}_2 = \begin{bmatrix} 0 & 1 & 1 \\ 1 & 1 & 1 \\ 1 & 2 & 1 \end{bmatrix}$$

$$\bar{C}_1 = \begin{bmatrix} -6 & 4 \\ 1 & -1 \\ 0 & 0 \end{bmatrix} \quad \bar{C}_2 = \begin{bmatrix} 5/3 & 17/3 & 1/3 \\ 0 & -1 & 0 \\ 0 & 0 & 0 \end{bmatrix}$$

The right characteristic λ-matrix of $(\bar{A}_2, \bar{B}_2)$ is

$$D_{r2}(\lambda) = \begin{bmatrix} \lambda^2-\lambda-2 & -2\lambda-2 & 1 \\ 0 & \lambda+1 & -1 \\ 0 & 0 & 1 \end{bmatrix}$$

and the left characteristic λ-matrix of $(\bar{A}_1, \bar{C}_1)$ is

$$D_{\ell 1}(\lambda) = \begin{bmatrix} \lambda-2 & -18 & 0 \\ 0.5 & \lambda+4 & 0 \\ 0 & 0 & 1 \end{bmatrix}$$

Therefore from Theorem 4.12, we have

$$D_r(\lambda) = D_{r2}(\lambda)U_{r\ell}(\lambda)D_{\ell 1}(\lambda)$$

where $U_{r\ell}(\lambda)$ is unimodular and can be found as

$$U_{r\ell}(\lambda) = \begin{bmatrix} 0 & 2 & 0 \\ -0.5 & \lambda-2 & 0 \\ 0 & 0 & 1 \end{bmatrix}$$

$D_r(\lambda)$ can also be factored as

$$D_r(\lambda) = D_{r2}(\lambda)\bar{D}_{\ell 1}(\lambda)$$

where

$$\bar{D}_{\ell 1}(\lambda) = U_{r\ell}(\lambda)D_{\ell 1}(\lambda) = \begin{bmatrix} 1 & 2\lambda+8 & 0 \\ 0 & \lambda^2+2\lambda+1 & 0 \\ 0 & 0 & 1 \end{bmatrix}$$

which is not a canonical λ-matrix

Example 4.2

Consider a reachable pair (A,B) of a continuous-time system:

$$A = \begin{bmatrix} -1.0 & -6.0 & 11.0 & -32.0 & 63.0 \\ -1.0 & -2.0 & 3.0 & -11.0 & 24.0 \\ 2.0 & -4.0 & 10.0 & -18.0 & 37.0 \\ 1.0 & -2.0 & 4.0 & -7.0 & 12.0 \\ 0.0 & 0.0 & -0.5 & 1.0 & -3.0 \end{bmatrix}$$

and

$$B = \begin{bmatrix} -2.0 & 1.0 & 0.0 \\ 1.0 & -2.0 & -3.0 \\ 1.0 & 1.0 & 3.0 \\ 2.0 & 1.0 & 4.0 \\ 1.0 & 0.0 & 1.0 \end{bmatrix}$$

The right characteristic λ-matrix is found to be

$$D_r(\lambda) = \begin{bmatrix} \lambda^3+3.6667\lambda^2-0.6111\lambda-4.0556 & \lambda^2-0.3333\lambda-0.6667 & -1.0 \\ -0.5556\lambda-9.1667 & \lambda^2-0.6667\lambda-2.0000 & -2.0 \\ 0.0 & 0.0 & 1.0 \end{bmatrix}$$

From Lemma 2.1, a minimal realization quadruple of $D_r^{-1}(A)$ is (A, B, C_r, D_r), where

$$C_r = \begin{bmatrix} 0.29333 & -0.64000 & 1.38667 & -2.96000 & 5.76000 \\ -1.04444 & 2.06667 & -4.42222 & 9.60000 & -18.33333 \\ 0.00000 & 0.00000 & 0.00000 & 0.00000 & 0.00000 \end{bmatrix}$$

$$D_r = \begin{bmatrix} 0.0 & 0.0 & 0.0 \\ 0.0 & 0.0 & 0.0 \\ 0.0 & 0.0 & 1.0 \end{bmatrix}$$

Selecting $r_0 = -\infty$, $r_1 = -3$, $r_2 = 0$, and $r_3 = \infty$, we have the following shifted matrix sign functions:

$$\text{Sign}_{(r_0)}(A) = I_5$$

$$\text{Sign}_{(r_1)}(A) = \begin{bmatrix} 1.0 & -4.0 & 8.0 & -20.0 & 40.0 \\ 0.0 & -1.0 & 4.0 & -10.0 & 20.0 \\ 0.0 & 0.0 & 1.0 & 0.0 & 0.0 \\ 0.0 & 0.0 & 0.0 & 1.0 & 0.0 \\ 0.0 & 0.0 & 0.0 & 0.0 & 1.0 \end{bmatrix}$$

$$\text{Sign}_{(r_2)}(A) = \begin{bmatrix} 1.0 & -4.0 & 10.0 & -24.0 & 48.0 \\ 0.0 & -1.0 & 0.0 & -20.0 & 4.0 \\ 0.0 & 0.0 & -1.0 & 4.0 & -8.0 \\ 0.0 & 0.0 & 0.0 & 1.0 & -4.0 \\ 0.0 & 0.0 & 0.0 & 0.0 & -1.0 \end{bmatrix}$$

$$\text{Sign}_{(r_3)}(A) = -I_5$$

From Theorem 4.19, we have

$$V_1 = [0.0 \quad 2.0 \quad -4.0 \quad 10.0 \quad -20.0]$$

$$V_2 = \begin{bmatrix} 0.0 & 0.0 & -1.0 & 2.0 & -4.0 \\ 0.0 & 0.0 & 0.0 & 0.0 & 2.0 \end{bmatrix}$$

$$V_3 = \begin{bmatrix} 1.0 & -2.0 & 5.0 & -12.0 & 24.0 \\ 0.0 & 0.0 & 0.0 & -1.0 & 2.0 \end{bmatrix}$$

Define
$$M_V = \begin{bmatrix} V_1 \\ V_2 \\ V_3 \end{bmatrix}$$

we have the block diagonalized minimal realization quadruple of $D_r^{-1}(A)$:

$$A_D = M_V A M_V^{-1} = \text{Block diag} \left[(-4.0), \begin{pmatrix} 0.0 & -0.5 \\ 1.0 & -1.0 \end{pmatrix}, \begin{pmatrix} -1.0 & 4.0 \\ -1.0 & 3.0 \end{pmatrix} \right]$$

$$B_D = M_V B = \begin{bmatrix} -2.0 & 2.0 & 2.0 \\ -1.0 & 1.0 & 1.0 \\ 1.0 & 0.0 & 2.0 \\ 1.0 & -2.0 & -3.0 \\ 0.0 & -1.0 & -2.0 \end{bmatrix}$$

$$C_D = C_r M_V^{-1} = \begin{bmatrix} -0.02667 & \vert & 0.18667 & -0.08003 & \vert & 0.29333 & -0.45333 \\ -0.01111 & \vert & -0.75556 & 0.13333 & \vert & -1.04444 & 1.31111 \\ 0.00000 & \vert & 0.00000 & 0.00000 & \vert & 0.00000 & 0.00000 \end{bmatrix}$$

$$D_D = D_r$$

From Theorem 4.13, we have a complete set of canonical left divisors $\{D_{\ell 1}(\lambda),$ $D_{\ell 2}(\lambda), D_{\ell 3}(\lambda)\}$:

$$D_{\ell 1}(\lambda) = \begin{bmatrix} \lambda+4.0 & 1.0 & 1.0 \\ 0.0 & 1.0 & 0.0 \\ 0.0 & 0.0 & 1.0 \end{bmatrix}$$

$$D_{\ell 2}(\lambda) = \begin{bmatrix} \lambda+1.5 & -0.5 & -1.0 \\ 2.5 & \lambda-0.5 & -2.0 \\ 0.0 & 0.0 & 1.0 \end{bmatrix}$$

and

$$D_{\ell 3}(\lambda) = \begin{bmatrix} \lambda-1.0 & 0.0 & -1.0 \\ -1.0 & \lambda-1.0 & -2.0 \\ 0.0 & 0.0 & 1.0 \end{bmatrix} ;$$

and a complete set of canonical right divisors $\{D_{r1}(\lambda), D_{r2}(\lambda), D_{r3}(\lambda)\}$:

$$D_{r1}(\lambda) = \begin{bmatrix} \lambda+4.00000 & 0.0 & 0.0 \\ 0.41667 & 1.0 & 0.0 \\ 0.00000 & 0.0 & 1.0 \end{bmatrix}$$

$$D_{r2}(\lambda) = \begin{bmatrix} \lambda-0.58394 & -0.25000 & 0.0 \\ 5.69761 & \lambda+1.58394 & 0.0 \\ 0.0 & 0.0 & 1.0 \end{bmatrix}$$

and

$$D_{r3}(\lambda) = \begin{bmatrix} \lambda+0.16667 & 0.20000 & 0.0 \\ -6.80556 & \lambda-2.16667 & 0.0 \\ 0.0 & 0.0 & 1.0 \end{bmatrix}$$

To perform the spectral factorization, using Theorem 4.20, we have

$$T = \begin{bmatrix} S^+_{(r_2,r_3)} \\ \bar{V}_{(r_2,r_3)} \end{bmatrix} = \begin{bmatrix} 1.00000 & -2.00000 & 4.00000 & 2.00000 & 0.00000 \\ 0.00000 & -0.16667 & 0.33333 & 0.16667 & 0.00000 \\ 0.00000 & 2.00000 & -5.00000 & 12.00000 & -24.00000 \\ 0.00000 & 1.00000 & 0.00000 & 1.00000 & -2.00000 \\ 0.00000 & 0.00000 & 0.00000 & 0.00000 & 2.00000 \end{bmatrix}$$

and a minimal realization quadruple (A_T, B_T, C_T, D_T) of $D_r^{-1}(\lambda)$ with A_T being block-triangularized:

$$A_T = TAT^{-1} = \begin{bmatrix} A_R & A_{RL} \\ 0_{3\times2} & A_L \end{bmatrix} = \begin{bmatrix} 11.0 & -100.0 & -8.4660 & 0.2652 & -7.8314 \\ 1.0 & -9.0 & 0.5000 & -0.1667 & -0.6667 \\ 0.0 & 0.0 & -3.2000 & -1.6000 & 0.5000 \\ 0.0 & 0.0 & -1.6000 & -0.8000 & 1.0000 \\ 0.0 & 0.0 & 0.2000 & 0.4000 & -1.0000 \end{bmatrix}$$

$$B_T = TB = \begin{bmatrix} B_R \\ B_L \end{bmatrix} = \begin{bmatrix} 4.0000 & 11.0000 & 26.0000 \\ 0.5000 & 0.83333 & 2.1667 \\ -3.0000 & 3.00000 & -3.0000 \\ 1.0000 & -1.0000 & -1.0000 \\ 2.0000 & 0.00000 & 2.0000 \end{bmatrix}$$

$$C_T = CT^{-1} = [C_R, C_L] = \begin{bmatrix} 2.9333 & -3.0667 & -0.2471 & 0.0703 & -0.1555 \\ -1.0444 & 11.2222 & 0.7970 & 0.2542 & 0.3517 \\ 0.0000 & 0.0000 & 0.0000 & 0.0000 & 0.0000 \end{bmatrix}$$

and

$$D_T = D_r$$

From Theorem 4.12, $D_r(\lambda)$ can be factored as

$$D_r(\lambda) = \bar{D}_{r2}(\lambda)U_{r\ell}(\lambda)\bar{D}_{\ell 1}(\lambda)$$

where $\bar{D}_{r2}(\lambda)$ is the right characteristic λ-matrix of (A_L, B_L):

$$\bar{D}_{r2}(\lambda) = \begin{bmatrix} \lambda^2+3.5\lambda+4.25 & 0.8\lambda+0.7 & -1.0 \\ 6.25 & \lambda+1.5 & -2.0 \\ 0.00 & 1.0 & 1.0 \end{bmatrix}$$

$\bar{D}_{\ell 1}(\lambda)$ is the left characteristic λ-matrix of (A_R, C_R):

$$\bar{D}_{\ell 1}(\lambda) = \begin{bmatrix} \lambda+0.1667 & 0.2000 & 0.0 \\ -6.8056 & \lambda-2.1667 & 0.0 \\ 0.0000 & 0.0000 & 1.0 \end{bmatrix}$$

and $U_{r\ell}(\lambda)$ is a unimodular λ-matrix:

$$U_{r\ell}(\lambda) = I_3$$

To factor $\bar{D}_{r2}(\lambda)$, we find a minimal realization quadruple $(A_L, B_L, \bar{C}_L, \bar{D}_L)$ of $D_{r2}^{-1}(\lambda)$:

$$A_L = \begin{bmatrix} -3.20 & -1.60 & 0.50 \\ -1.60 & -0.80 & 1.00 \\ 0.20 & 0.40 & -1.00 \end{bmatrix}$$

$$B_L = \begin{bmatrix} -3.00 & 3.00 & 3.00 \\ 1.00 & -1.00 & -1.00 \\ 2.00 & 0.00 & 2.00 \end{bmatrix}$$

$$\bar{C}_L = \begin{bmatrix} 0.04 & 0.12 & 0.00 \\ 0.30 & -0.10 & 0.50 \\ 0.00 & 0.00 & 0.00 \end{bmatrix}$$

$$\bar{D}_L = D_r$$

Selecting $r_0 = -\infty$, $r_1 = -3$ and $r_2 = \infty$, we have the following shifted matrix sign functions

$$\text{Sign}_{(r_0)}(A_L) = I_3$$

$$\text{Sign}_{(r_1)}(A_L) = \begin{bmatrix} -0.6 & -0.8 & 0.0 \\ -0.8 & 0.6 & 0.0 \\ 0.0 & 0.0 & 1.0 \end{bmatrix}$$

$$\text{Sign}_{(r_2)}(A_L) = -I_3$$

Again, from Theorem 4.20, we have

$$T_L = \begin{bmatrix} S^+_{L(r_1,r_2)} \\ \\ V_{L(r_1,r_2)} \end{bmatrix} = \begin{bmatrix} 1.0 & -2.0 & 0.0 \\ 0.0 & 0.0 & 1.0 \\ \hline 0.8 & 0.4 & 0.0 \end{bmatrix}$$

and a minimal realization quadruple $(A_{LT}, B_{LT}, C_{LT}, D_{LT})$ of $\bar{D}^{-1}_{r2}(\lambda)$ with A_{LT} being block-triangularized:

$$A_{LT} = T_L A T_L^{-1} = \begin{bmatrix} A_{LR} & A_{LRL} \\ \\ 0 & A_{LL} \end{bmatrix} = \begin{bmatrix} 0.0 & -0.25 & 0.0 \\ 0.2 & -1.0 & 0.0 \\ \hline 0.0 & 0.0 & -4.0 \end{bmatrix}$$

$$B_{LT} = T_L B = \begin{bmatrix} B_{LR} \\ \\ B_{LL} \end{bmatrix} = \begin{bmatrix} -5.0 & 5.0 & 5.0 \\ 2.0 & 0.0 & 2.0 \\ \hline -2.0 & 2.0 & 2.0 \end{bmatrix}$$

$$C_{LT} = C T_L^{-1} = [C_{LR} \ C_{LL}] = \begin{bmatrix} 0.04 & 0.0 & 1.0 \\ 0.10 & 0.50 & 0.25 \\ 0.00 & 0.00 & 0.00 \end{bmatrix}$$

$$D_{LT} = D_r$$

Now, from Theorem 4.12 we have

$$\bar{\bar{D}}_{r2}(\lambda) = \bar{\bar{D}}_{r2}(\lambda)\bar{U}_{r\ell}(\lambda)\bar{\bar{D}}_{\ell 1}(\lambda)$$

where $\bar{\bar{D}}_{r2}(\lambda)$ is the right characteristic λ-matrix of (A_{LL}, B_{LL}):

$$\bar{\bar{D}}_{r2}(\lambda) = \begin{bmatrix} \lambda+4.00 & 1.00 & 1.00 \\ 0.00 & 1.00 & 0.00 \\ 0.00 & 0.00 & 1.00 \end{bmatrix}$$

$\bar{\bar{D}}_{\ell 1}(\lambda)$ is the left characteristic λ-matrix of (A_{LR}, C_{LR}):

$$\bar{\bar{D}}_{\ell 1}(\lambda) = \begin{bmatrix} \lambda-0.50 & -0.20 & 0.00 \\ 6.25 & \lambda+1.50 & 0.00 \\ 0.00 & 0.00 & 0.00 \end{bmatrix}$$

and $\bar{U}_{r\ell}(\lambda)$ is a unimodular λ-matrix:

$$\bar{U}_{r\ell}(\lambda) = \begin{bmatrix} 1.0 & 0.0 & 0.0 \\ 0.0 & 1.0 & -2.0 \\ 0.0 & 0.0 & 1.0 \end{bmatrix}$$

Combining these two parts of spectral factorization, we have

$$D_r(\lambda) = \begin{bmatrix} \lambda+4.00 & 1.00 & 1.00 \\ 0.00 & 1.00 & 0.00 \\ 0.00 & 0.00 & 1.00 \end{bmatrix} \begin{bmatrix} 1.0 & 0.0 & 0.0 \\ 0.0 & 1.0 & -2.0 \\ 0.0 & 0.0 & 1.0 \end{bmatrix} \times$$

$$\begin{bmatrix} \lambda-0.50 & -0.20 & 0.00 \\ 6.25 & \lambda+1.50 & 0.00 \\ 0.00 & 0.00 & 1.00 \end{bmatrix} \begin{bmatrix} \lambda+0.1667 & 0.2000 & 0.0 \\ -6.8056 & \lambda-2.1667 & 0.0 \\ 0.000 & 0.0000 & 1.0 \end{bmatrix}$$

CHAPTER V   FEEDBACK CONTROL OF MULTIVARIABLE SYSTEMS

Feedback design for multivariable systems has attracted many researchers [5-16,62-78] in the fields of system theory and control engineering. Probably the most successful design methods are state-feedback control in the time domain [5,63-71], and the inverse Nyquist array and the root loci design in the frequency domain [8,72-78]. In this chapter, we shall discuss the feedback control of multivariable systems employing the notions of the characteristic λ-matrices and their divisors for constructing the state-feedback control laws. These new approaches permit a deeper insight into some structural aspects of feedback control. In general, the design schemes presented in this chapter provide methods for constructing state feedback controllers by dealing with the assignments of the denominator of the closed-loop matrix fraction description. Therefore, advantages of both the time domain and the frequency domain design can be achieved via these design methods.

Some properties of linear state-feedback controls are set out in Section 5.1. In Section 5.2, methods using characteristic and non-characteristic λ-matrix assignments to construct the state-feedback controller are developed. Section 5.3 is devoted to deriving the latent structure assignments which take the left/right generalized latent vectors of the closed-loop right characteristic λ-matrices as the design guideline. The divisor assignment and decoupling design approaches are discussed in Section 5.4.

5.1 Linear State-Feedback Controls and Properties

Consider an m-input, p-output linear time-invariant system described by

$$\lambda X(t) = AX(t) + Bu(t) \tag{5.1a}$$

$$y(t) = CX(t) + Du(t) \tag{5.1b}$$

where $X(t) \in C^n$, $u(t) \in C^m$, and $y(t) \in C^p$ are state, input, and output vectors, respectively; A, B, C, and D are matrices of appropriate dimensions; λ is an operator, and t is an independent variable. For continuous time systems, λ is a differential operator with $t \in R$. For discrete time systems, λ is a forward shift operator with $t \in Z$. Assume that the system in Eq. (5.1) is reachable. The linear state-feedback control law can be described as:

$$u(t) = -FX(t) + G_F r(t) \tag{5.2}$$

where F, G_F, and r(t) are the feedback gain, input gain, and reference input, respectively. The important invariance property of the Kronecker indices under a state-feedback control is as follows.

<u>Lemma 5.1</u>   Let the state-feedback control law be Eq. (5.2), and let $D_{rc}(\lambda)$ be the right characteristic λ-matrix of the closed-loop system. Then, $\partial_{ci}[D_{rc}(\lambda)]$ $= \partial_{ci}[D_r(\lambda)] = \kappa_i, i=1,\ldots,m$, where $D_r(\lambda)$ is the right characteristic λ-matrix of the open-loop system, $\kappa_i, i=1,\ldots,m$ are the Kronecker indices, and $\partial_{ci}[D_{rc}(\lambda)]$ denotes the ith column degree of $D_{rc}(\lambda)$.

<u>Proof:</u>

Refer to Popov [35].   ■

Lemma 5.1 gives the structure of all possible right characteristic λ-matrices of the closed-loop systems with linear state-feedback control laws. In other words, we may only assign the closed-loop right characteristic λ-matrix having the same Kronecker indices as the open-loop right characteristic λ-matrix. However, this structural limitation does not restrict the freedom of choosing the closed-loop eigenvalues. Thus, the dynamics of the closed-loop system still can be controlled by assigning appropriate eigenvalues or poles.

From Lemma 5.1, we obtain the results below.

Lemma 5.2   Let $(A,B,\bar{C},\bar{D})$ be a minimal realization of $D_r^{-1}(\lambda)$, where $D_r(\lambda)$ is the right characteristic λ-matrix of the reachable pair (A,B).   Let $D_{rc}(\lambda)$ be the desired right characteristic λ-matrix of the closed-loop system under the state-feedback control law in Eq. (5.2).   Then, $(A-BF,BG_F,\bar{C},\bar{D}G_F)$ is a minimal realization of $D_{rc}^{-1}(\lambda)$, where $G_F = D_{rh}D_{rhc}^{-1}$, D_{rhc} is the leading-column coefficient matrix of $D_{rc}(\lambda)$, and D_{rh} is that of $D_r(\lambda)$.

Proof:

From Lemma 2.1, we have

$$D_r^{-1}(\lambda) = \bar{C}(\lambda I_n-A)^{-1}B+\bar{D} = \psi_r^T(0)(\lambda I_n-A_c)^{-1}B_c+\bar{D} \tag{5.3}$$

where $\bar{C} = \psi_r^T(0)T_c$; $\bar{D} = (I_m-\psi_r^T(0)\psi_r(0))D_{rh}^{-1}$; $A_c = T_cAT_c^{-1}$; $B_c = T_cB$. (A_c,B_c) is the controller canonical form of (A,B).   Let

$$F_c \overset{\Delta}{=} FT_c^{-1} = D_{rh}\bar{F}_c \tag{5.4}$$

we obtain

$$T_c(A-BF)T_c^{-1} = A_c-B_cF_c = A_c-E_{bc}\bar{F}_c \overset{\Delta}{=} \bar{A}_c \tag{5.5a}$$

$$T_cBG_F = B_cD_{rh}D_{rhc}^{-1} = E_{bc}D_{rhc}^{-1} \overset{\Delta}{=} \bar{B}_c \tag{5.5b}$$

where $\bar{A}_c$ and $\bar{B}_c$ are the controller canonical forms of the closed-loop system. From Lemma 5.1, the Kronecker indices of $(A-BF,BG_F)$ are the same as those of (A,B), thus we obtain the minimal realization of the inverse of the right characteristic λ-matrix of $(A-BF,BG_F)$ as

$$D_{rc}^{-1}(\lambda) = \psi_r^T(0)(\lambda I_n - \bar{A}_c)^{-1}\bar{B}_c + \bar{D}_c = \bar{C}(\lambda I_n - (A-BF))^{-1}BG_F + \bar{D}G_F \qquad (5.6)$$

where $\bar{D}_c = (I_m - \psi_r^T(0)\psi_r(0))D_{rhc}^{-1} = \bar{D}G_F$. Thus, $(A-BF, BG_F, \bar{C}, \bar{D}G_F)$ is the minimal realization of $D_{rc}^{-1}(\lambda)$. ∎

Lemma 5.2 reveals the state space structure of the closed-loop right characteristics λ-matrix under linear state-feedback controls. From Lemma 5.2, we have the following results on the structure of the feedback gain when some of the Kronecker indices are zero.

<u>Corollary 5.1</u> Let $\kappa \triangleq \{\kappa_1,\ldots,\kappa_m\}$ be the Kronecker indices of (A,B). If $\kappa_i = 0$ for some $1 \leq i \leq m$, then the ith rows of $\bar{F}_c$ defined in Eq. (5.4), can be chosen to be null, and therefore the ith rows of F_c and F are also null.

<u>Proof:</u>

From Eq. (5.5a), we have $\bar{A}_c = A_c - E_{bc}\bar{F}_c$. If $\kappa_i = 0$, the ith column of E_{bc} is null. Thus, the ith row of $E_{bc}\bar{F}_c$ is null, and we can choose $(\bar{F}_c)_i = 0$. Since $F = F_c T_c = D_{rh}\bar{F}_c T_c$ and $(D_{rh})_{ij} = 0$ for $j \neq i$ and $\kappa_i = 0$, the ith rows of $F_c = D_{rh}\bar{F}_c$ and F_c are both null. ∎

In the following sections of this chapter, we discuss various schemes for constructing linear state-feedback controls laws for different design purposes.

5.2 λ-matrix Assignment

In this section, we discuss methods to construct the linear state-feedback control laws from the desired denominator of the closed-loop RMFD. One simple way to assign the denominator of the closed-loop RMFD is to select a column-reduced canonical λ-matrix which has the same Kronecker indices as those of the open loop right characteristic λ-matrix. This approach is referred to as the right characteristic λ-matrix assignment. For this scheme, the numerator λ-matrix of the closed-loop canonical RMFD is to be the same as that of the open loop canonical RMFD. A second way to approach the problem is to assign the denominator of the closed-loop RMFD as a non-canonical column-reduced λ-matrix having the same Kronecker indices as those of the open loop right characteristic λ-matrix. This approach is referred to as the column-reduced λ-matrix assignment. The important feature of this scheme is that the numerator λ-matrix of the closed-loop canonical RMFD may not be the same as that of the open-loop canonical RMFD. This result is different from the situation in SISO systems using linear state-feedback controls.

<u>Theorem 5.1</u>   (Right Characteristic λ-matrix Assignment)

If the desired closed-loop right characteristic λ-matrix is given by $D_{rc}(\lambda)$, and D_{rhc} is the leading-column matrix of $D_{rc}(\lambda)$, then the feedback gain and input gain become

$$F = F_c T_c \qquad\qquad (5.7a)$$

$$G_F = D_{rh} D_{rhc}^{-1} \qquad\qquad (5.7b)$$

where F_c can be determined by

$$F_c \psi_r(\lambda) = G_F D_{rc}(\lambda) - D_r(\lambda) \qquad\qquad (5.7c)$$

<u>Proof</u>:

From Lemma 5.2, we obtain

$$\bar{B}_c D_{rc}(\lambda) = (\lambda I_n - \bar{A}_c)\psi_r(\lambda) = (\lambda I_n - A_c + E_{bc}\bar{F}_c)\psi_r(\lambda)$$

$$= (\lambda I_n - A_c)\psi_r(\lambda) + E_{bc}\bar{F}_c\psi_r(\lambda) = B_c D_r(\lambda) + E_{bc}\bar{F}_c\psi_r(\lambda)$$

128

Therefore,

$$B_c(G_F D_{rc}(\lambda) - D_r(\lambda)) = E_{bc}\bar{F}_c \psi_r(\lambda)$$

Since $B_c = E_{bc}D_{rh}^{-1}$ and the ith row of $\bar{F}_c$ is zero if $\kappa_i = 0$, we have $D_{rh}^{-1}(G_F D_{rc}(\lambda) - D_r(\lambda)) = \bar{F}_c \psi_r(\lambda)$.

Thus, the result of Eq. (5.7c) follows. ∎

The closed-loop canonical RMFD [79,80] can be found as follows.

Corollary 5.2   Let the desired closed-loop right characteristic λ-matrix be $D_{rc}(\lambda)$ and F and G_F be determined in Theorem 5.1.   Then, the closed-loop canonical RMFD with $D = 0_{p \times m}$ is given by

$$G_c(\lambda) = C(\lambda I_n - (A - BF))^{-1} BG_F = N_r(\lambda) D_{rc}^{-1}(\lambda) \tag{5.8}$$

Proof:

$$G_c(\lambda) = C(\lambda I_n - (A - BF))^{-1} BG_F = C_c(\lambda I_n - \bar{A}_c)^{-1}\bar{B}_c$$

$$= C_c \psi_r(\lambda) D_{rc}^{-1}(\lambda) = N_r(\lambda) D_{rc}^{-1}(\lambda) \qquad ∎$$

If arbitrary F and G_F are assigned in Eq. (5.2), the closed-loop RMFD, which may not be the canonical RMFD, can be described as follows.

Theorem 5.2   Let $D = 0_{p \times m}$ and the state-feedback control law be shown in Eq. (5.2) with the ith row of F being null if $\kappa_i = 0$ and G_F being nonsingular.   Then, the RMFD of the closed-loop system can be expressed as

$$G_c(\lambda) = C(\lambda I_n - (A-BF))^{-1}BG_F = N_r(\lambda)\hat{D}_{rc}^{-1}(\lambda) \tag{5.9a}$$

where $\hat{D}_{rc}(\lambda)$ is a column-reduced λ-matrix defined by

$$\hat{D}_{rc}(\lambda) = G_F^{-1}[FT_c^{-1}\psi_r(\lambda)+D_r(\lambda)] \tag{5.9b}$$

Proof:

$$G_c(\lambda) = C(\lambda I_n - (A-BF))^{-1}BG_F = C_c(\lambda I_n - \bar{A}_c)^{-1}\bar{B}_c$$

where

$$\bar{A}_c = A_c - B_c FT_c^{-1} \text{ and } \bar{B}_c = T_c BG_F = E_{bc}D_{rh}^{-1}G_F$$

For the arbitrary matrices F and G_F specified in Theorem 5.2, $\bar{A}_c$ and $\bar{B}_c$ are in controller forms but not necessary in canonical controller forms. Therefore, from Lemma 3.2, we have

$$\hat{D}_{rc}^{-1}(\lambda) = \psi_r^T(0)(\lambda I_n - \bar{A}_c)^{-1}\bar{B}_c + \bar{D}_c$$

where $\bar{D}_c = (I_m - \psi_r^T(0)\psi_r(0))D_{rh}^{-1}G_F$. It can be easily shown that

$$\bar{B}_c\hat{D}_{rc}(\lambda) = (\lambda I_n - \bar{A}_c)\psi_r(\lambda)$$

whence

$$B_c G_F\hat{D}_{rc}(\lambda) = (\lambda I_n - A_c + B_c FT_c^{-1})\psi_r(\lambda) = (\lambda I_n - A_c)\psi_r(\lambda) + B_c FT_c^{-1}\psi_r(\lambda)$$

$$= B_c D_r(\lambda) + B_c FT_c^{-1}\psi_r(\lambda)$$

Since the ith row of F is null if $\kappa_i = 0$, we have the result in Eq. (5.9b), from which

$$G_F \hat{D}_{rc}(\lambda) = D_r(\lambda) + FT_c^{-1}\psi_r(\lambda)$$

Thus, the closed-loop RMFD becomes

$$G_c(\lambda) = C_c(\lambda I_n - \bar{A}_c)^{-1}\bar{B}_c = C_c\psi_r(\lambda)\hat{D}_{rc}^{-1}(\lambda) = N_r(\lambda)\hat{D}_{rc}^{-1}(\lambda) \qquad \blacksquare$$

Remark

Theorem 5.2 reveals the fact that the closed-loop RMFD can be written as $N_r(\lambda)\hat{D}_{rc}^{-1}(\lambda)$, where the RMFD may not be canonical even if $N_r(\lambda)$ is the numerator of the open-loop <u>canonical</u> RMFD and $\hat{D}_{rc}(\lambda)$ is column-reduced λ-matrix determined by $D_r(\lambda)$, F and G_F. If the <u>canonical</u> RMFD of the closed-loop system is found to be $G_c(\lambda) = N_{rc}(\lambda)D_{rc}^{-1}(\lambda)$, we can easily prove that $N_{rc}(\lambda) = N_r(\lambda)U(\lambda)$ and $D_{rc}(\lambda) = \hat{D}_{rc}(\lambda)U(\lambda)$ where $U(\lambda)$ is a unimodular matrix.

From Theorems 5.1 and 5.2, we observe that the non-canonical column-reduced λ-matrix assignment can be accomplished as follows:

Theorem 5.3   (Column-reduced λ-matrix Assignment)

Let $\hat{D}_{rc}(\lambda)$ be a given column-reduced λ-matrix with column degrees $(\kappa_1,\ldots,\kappa_m)$, which is the set of Kronecker indices of the open loop system in Eq. (5.1). $\hat{D}_{rhc}$ is the leading-column matrix of $\hat{D}_{rc}(\lambda)$. We choose the feedback gain and input gain to be

$$F = F_c T_c \qquad (5.10a)$$

$$G_F = D_{rh}\hat{D}_{rhc}^{-1} \qquad (5.10b)$$

where F_c can be determined by

$$F_c\psi_r(\lambda) = G_F\hat{D}_{rc}(\lambda) - D_r(\lambda) \qquad (5.10c)$$

Then, the closed-loop RMFD with $D = 0_{p \times m}$ is given by

$$\hat{G}_c(\lambda) = C(\lambda I_n - (A-BF))^{-1} BG_F = N_r(\lambda) \hat{D}_{rc}^{-1}(\lambda) \qquad (5.11)$$

Proof:

    This follows directly from Theorems 5.1 and 5.2.   ■

Example 5.1

    Given a 3-input 2-output continuous time system

$$\lambda X(t) = AX(t) + Bu(t)$$
$$y(t) = CX(t) + Du(t)$$

where

$$A = \begin{bmatrix} 2.2 & 1.6 & 4.0 & 3.0 & 1.0 \\ -6.4 & -4.2 & -8.0 & -6.0 & -2.0 \\ -1.6 & -0.8 & -3.0 & -2.0 & -1.0 \\ 3.2 & 1.6 & 4.0 & 3.0 & 2.0 \\ 9.6 & 4.8 & 12.0 & 6.0 & 2.0 \end{bmatrix}$$

$$B = \begin{bmatrix} -2.0 & 1.0 & 0.0 \\ -1.0 & -2.0 & -5.0 \\ 3.0 & -1.0 & 1.0 \\ -2.0 & 3.0 & 4.0 \\ 1.0 & -2.0 & -3.0 \end{bmatrix}$$

$$C = \begin{bmatrix} 3.6 & 3.8 & 5.0 & 4.0 & 0.0 \\ 5.4 & 2.2 & 5.0 & 3.0 & -1.0 \end{bmatrix}$$

$$D = 0_{2 \times 3}$$

Using the minimal nice selection algorithm in Section 2.3, the Kronecker indices are found to be $\kappa_1 = 3$, $\kappa_2 = 2$ and $\kappa_3 = 0$. Note that the input matrix B contains a dependent vector, but the system is reachable. From Eq. (2.16), we have the canonical RMFD

$$G(\lambda) = N_r(\lambda)D_r^{-1}(\lambda)$$

where

$$N_r(\lambda) = \begin{bmatrix} -4\lambda^2+13\lambda-13 & 7\lambda-12 & 0 \\ -5\lambda^2+10\lambda+3 & 12\lambda+7 & 0 \end{bmatrix}$$

and

$$D_r(\lambda) = \begin{bmatrix} \lambda^3-2\lambda^2-\lambda+2 & -\lambda^2+\lambda+2 & -1 \\ 0 & \lambda^2+2\lambda+1 & -2 \\ 0 & 0 & 1 \end{bmatrix}$$

The open loop poles are $-1,-1,-1,1,2$. We shall construct linear state-feedback control laws using both the right characteristic λ-matrix assignment and the column-reduced λ-matrix assignment as follows:

(1) Right characteristic λ-matrix assignment

Let the desired closed-loop right characteristic λ-matrix be

$$D_{rc}(\lambda) = \begin{bmatrix} \lambda^3+6\lambda^2+22\lambda+30 & -\lambda^2-7\lambda-12 & 0 \\ -10\lambda-22 & \lambda^2+7\lambda+12 & 0 \\ 0 & 0 & 1 \end{bmatrix}$$

which is a column-reduced canonical λ-matrix with the leading-column coefficient matrix

$$D_{rhc} = \begin{bmatrix} 1 & -1 & 0 \\ 0 & 1 & 0 \\ 0 & 0 & 1 \end{bmatrix}$$

From Theorem 5.1, we have the feedback gain matrix

$$F = - \begin{bmatrix} -12.1 & -12.3 & -25.5833 & -26.4167 & -20.5833 \\ -25.6 & -6.8 & -22.0000 & -1.0000 & 6.0000 \\ 0.0 & 0.0 & 0.0000 & 0.0000 & 0.0000 \end{bmatrix}$$

133

and the input gain matrix

$$G_F = \begin{bmatrix} 1.0 & 0.0 & -1.0 \\ 0.0 & 1.0 & -2.0 \\ 0.0 & 0.0 & 1.0 \end{bmatrix}$$

The linear state-feedback control law is given by

$$u(t) = -FX(t)+G_F r(t)$$

The closed-loop canonical RMFD can be found as

$$G_c(\lambda) = N_r(\lambda)D_{rc}^{-1}(\lambda)$$

The closed loop poles are -2,-2,-2,-3,-4.

(2) Column-reduced λ-matrix assignment

Let the desired denominator of the closed loop RMFD be

$$\hat{D}_{rc}(\lambda) = \begin{bmatrix} \lambda^3+4\lambda^2+8\lambda+6 & -\lambda^2-7\lambda-12 & 0 \\ 2\lambda^2+4\lambda+2 & \lambda^2+7\lambda+12 & 0 \\ 0 & 0 & 1 \end{bmatrix}$$

The leading-column coefficient matrix is

$$\hat{D}_{rhc} = \begin{bmatrix} 1 & -1 & 0 \\ 0 & 1 & 0 \\ 0 & 0 & 1 \end{bmatrix}$$

which is nonsingular. Thus, $\hat{D}_{rc}(\lambda)$ is column-reduced with column degrees $\partial_{c1}(\hat{D}_{rc}(\lambda)) = 3$, $\partial_{c2}(\hat{D}_{rc}(\lambda)) = 2$ and $\partial_{c3}(\hat{D}_{rc}(\lambda)) = 0$. But

$$D_{rhc}^{-1}\hat{D}_{rc}(\lambda) = \begin{bmatrix} \lambda^3+6\lambda^2+12\lambda+8 & 0 & 0 \\ 2\lambda^2+4\lambda+2 & \lambda^2+7\lambda+12 & 0 \\ 0 & 0 & 1 \end{bmatrix}$$

and from Definition 3.1, $\hat{D}_{rc}(\lambda)$ is not a column-reduced canonical λ-matrix. From Theorem 5.3, we have the feedback gain

$$F = - \begin{bmatrix} -10.1 & -8.3 & -17.5833 & -14.4167 & -10.5833 \\ -27.6 & -10.8 & -30.0000 & -13.0000 & -4.0000 \\ 0.0 & -0.0 & 0.0000 & 0.0000 & 0.0000 \end{bmatrix}$$

and the input gain

$$G_F = \begin{bmatrix} 1.0 & 0.0 & -1.0 \\ 0.0 & 1.0 & -2.0 \\ 0.0 & 0.0 & 1.0 \end{bmatrix}$$

The linear state-feedback control law is

$$u(t) = -FX(t)+G_F r(t)$$

The closed-loop RMFD is given by

$$\hat{G}_c(\lambda) = N_r(\lambda)\hat{D}_{rc}^{-1}(\lambda)$$

Since the denominator of $\hat{G}_c(\lambda)$ is not a column-reduced canonical λ-matrix, $\hat{G}_c(\lambda)$ is not a canonical RMFD. From Theorem 3.4, we have the canonical RMFD of the closed-loop system:

$$G_c(\lambda) = N_{rc}(\lambda)D_{rc}^{-1}(\lambda)$$

where

$$N_{rc}(\lambda) = \begin{bmatrix} -4\lambda^2-\lambda+11 & 7\lambda-12 & 0 \\ -5\lambda^2-14\lambda-11 & 12\lambda+7 & 0 \end{bmatrix}$$

It is easy to check that

$$N_{rc}(\lambda) = N_r(\lambda)U_r(\lambda)$$

$$D_{rc}(\lambda) = \hat{D}_{rc}(\lambda)U_r(\lambda)$$

where $U_r(\lambda)$ is a unimodular λ-matrix

$$U_r(\lambda) = \begin{bmatrix} 1 & 0 & 0 \\ -2 & 1 & 0 \\ 0 & 0 & 1 \end{bmatrix}$$

The denominator of the canonical RMFD of the closed-loop system in this design example is the same as that in (1); however, the numerator is different (compare the numerator of the open loop canonical RMFD with $N_{rc}(\lambda)$).

5.3 Left and Right Latent Structure Assignment

In this section, we present an alternative method for constructing the linear state-feedback control law from the desired left or right latent structure of the closed-loop right characteristic λ-matrix. This approach is the counterpart of eigenstructure assignment for MIMO system design, which have been discussed by many authors [64,66]. An advantage of determining the feedback control law using the latent structure of the closed-loop right characteristic λ-matrix instead of the eigenstructure of the closed-loop system map, is to avoid dealing with the higher dimensional eigenvectors of the system map.

Using the relationships between the latent structures of the right characteristic λ-matrix of a reachable pair and the eigenstructures of the system map explored in Section 3.2, we derive the state-feedback control law via latent roots and latent vectors assignment as follows.

Theorem 5.4 (Left Latent Structure Assignment)

A linear state-feedback controller in Eq. (5.2) is required for controlling the m-input p-output reachable system in Eq. (5.1) with $D = 0_{p \times m}$ such that the closed-loop right characteristic λ-matrix has the assigned latent roots, $\{\bar{\lambda}_i, i=1,\ldots,k\}$, and left generalized latent vectors, $\{\bar{P}_{ij}, j=0,1,\ldots,$ $\ell_i-1, i=1,\ldots,k\}$, where ℓ_i is the length of Jordan chain corresponding to $\bar{\lambda}_i$ and $\sum\limits_{i=1}^{k} \ell_i=n$. Define

$$\bar{A}_J \triangleq \text{block diag}\{\bar{J}_i, i=1,\ldots,k\}\epsilon C^{n\times n} \qquad (5.11a)$$

where

$$\bar{J}_i \triangleq \begin{bmatrix} \bar{\lambda}_i & 1 & \cdot & \cdot \\ 1 & \bar{\lambda}_i & 1 & \cdot \\ \cdot & \cdot & \cdot & \cdot \\ \cdot & \cdot & \cdot & \bar{\lambda}_i \end{bmatrix} \epsilon^{\ell_i\times\ell_i}; \quad \bar{B}_J \triangleq \begin{bmatrix} \bar{B}_{J1} \\ \cdot \\ \cdot \\ \bar{B}_{Jk} \end{bmatrix} \epsilon^{n\times m}; \quad \bar{B}_{Ji} \triangleq \begin{bmatrix} \bar{P}_{i\ell_i-1}^T \\ \cdot \\ \cdot \\ \bar{P}_{i0}^T \end{bmatrix} \epsilon^{\ell_i\times m}$$

$$(5.11b)$$

Assume that $(\bar{A}_J,\bar{B}_J)$ is a reachable pair and the Kronecker indices of $(\bar{A}_J,\bar{B}_J)$ are equal to those of (A,B). Then, the desired feedback gain F and the input gain G_F can be determined by $F = F_c T_c$ and $G_F = D_{rh}D_{rhc}^{-1}$, where T_c is the transformation matrix which transforms A into the canonical controller form $A_c \triangleq T_c A T_c^{-1}$, and F_c is computed from $F_c\psi_r(\lambda) = G_F D_{rc}(\lambda)-D_r(\lambda)$. $D_{rc}(\lambda)$ is the right characteristic λ-matrix of $(\bar{A}_J,\bar{B}_J)$ with the leading-column matrix D_{rhc} and $D_r(\lambda)$ is the right characteristic λ-matrix of (A,B) with the leading-column matrix D_{rh}.

Proof:

Theorem 5.4 can be proved using Theorems 5.1 and 3.9. ∎

Corollary 5.3   With the linear state-feedback control law determined in Theorem 5.4, the left generalized eigenvectors of the closed-loop system map $\bar{A} = A-BF$ can be determined by

$$\bar{P}_{aij} = T_c^T \bar{T}_P \sum\limits_{k=0}^{j} \frac{1}{k!} \psi_r^{(k)}(\bar{\lambda}_i)D_{rhc}^T\bar{P}_{i(j-k)}, \quad 0\leq j\leq\ell_i-1, \; i=1,\ldots,k$$

where

$$\bar{T}_p \triangleq \{[\bar{T}_p]_{ij}\}, \quad 1 \le i \le m, \ 1 \le j \le m, \ \kappa_i \kappa_j \ne 0$$

$[\bar{T}_p]_{ii}$ = reversed upper triangular Toeplitz matrix with first column $[\hat{a}_{ii2}, \ldots, \hat{a}_{ii\kappa_i}, 1]^T$

$[\bar{T}_p]_{ji}$ = reversed upper triangular Toeplitz matrix with first column $[\hat{a}_{ij2}, \ldots, \hat{a}_{ij\kappa_i}, 0]^T$ if $\kappa_i \le \kappa_j$ or $[\hat{a}_{ij2}, \ldots, \hat{a}_{ij\kappa_j}, 0, \ldots, 0]^T$ if $\kappa_i > \kappa_j$.

$\hat{a}_{ijk}$, $k = 1, \ldots, \min(\kappa_i, \kappa_j)$, are the coefficients of the (i,j)th entry of $\bar{\delta}_r(\lambda)$, where $\bar{\delta}_r(\lambda) = D_{rhc}^{-1} D_{rc}(\lambda)$.

Proof:

   Corollary 5.3 can be proved using Theorem 5.4 and Corollary 3.3. ∎

Theorem 5.5   (Right Latent Structure Assignment)

   A linear state-feedback controller in Eq. (5.2) is required for controlling the reachable system in Eq. (5.1) with $D = 0_{p \times m}$ such that the closed-loop right characteristic λ-matrix has the assigned latent roots, $\{\bar{\lambda}_i, i = 1, \ldots, k\}$, and right generalized latent vectors, $\{\bar{q}_{ij}, j = 0, 1, \ldots, \ell_i - 1, i = 1, \ldots, k\}$, where ℓ_i is the length of Jordan chain corresponding to $\bar{\lambda}_i$ and $\sum_{i=1}^{k} \ell_i = n$. Define $\bar{A}_J \triangleq$ block diag$[\bar{J}_i, i = 1, \ldots, k] \in C^{\ell_i \times \ell_i}$, where $\bar{J}_i$ is defined in Theorem 5.4, and $\bar{C}_J \triangleq [\bar{C}_{J1}, \ldots, \bar{C}_{Jk}] \in C^{m \times n}$ where $\bar{C}_{Ji} = [\bar{q}_{i0}, \ldots, \bar{q}_{i\ell_i - 1}] \in C^{m \times \ell_i}$. Assume that $(\bar{A}_J, \bar{C}_J)$ is an observable pair. The observability indices of $(\bar{A}_J, \bar{C}_J)$ are the Kronecker indices of (A, B), and the left characteristic λ-matrix of $(\bar{A}_J, \bar{C}_J)$ is $\bar{D}_\ell(\lambda)$ with a leading-row matrix I_m. Then, the feedback gain, F, and the input gain, G_F, can be determined by $F = F_c T_c$ and $G_F = D_{rh} D_{rhc}^{-1}$ where T_c is the transformation matrix which transforms A into its canonical controller form, $A_c = T_c A T_c^{-1}$. F_c is computed from the following matrix equation:

$$F_c \psi_r(\lambda) = G_F D_{rc}(\lambda) - D_r(\lambda) \tag{5.12}$$

where $D_{rc}(\lambda) = D_{rhc}\bar{D}_\ell(\lambda)$, D_{rhc} is any square upper triangular constant matrix with diagonal elements all 1's and $D_r(\lambda)$ is the right characteristic λ-matrix of (A,B).

Proof:

From the assumption and Lemma 2.3, $\bar{D}_\ell^{-1}(\lambda)$ can be realized by

$$\bar{D}_\ell^{-1}(\lambda) = \hat{\bar{C}}_J(\lambda I_n - \bar{A}_J)^{-1}\hat{\bar{B}}_J + \hat{\bar{D}}_J \tag{5.13}$$

where $\hat{\bar{B}}_J = \bar{T}_{0J}\bar{\psi}_\ell^T(0)$; $\hat{\bar{D}}_J = \bar{D}_{\ell h}^{-1}[I_m - \bar{\psi}_\ell^T(0)]$; $\bar{\psi}_\ell(\lambda) = \psi_r^T(\lambda)$; $\bar{D}_{\ell h} = I_m$. $\bar{T}_{0J}$ is the transformation matrix which transforms $\bar{A}_J$ into its observer canonical form $\bar{A}_0 = \bar{T}_{0J}^{-1}\bar{A}_J\bar{T}_{0J}$.

Since $\bar{D}_{\ell h} = I_m$, the (i,j)th entry of $\bar{D}_\ell(\lambda))_{ij}$, has degree $\min(\nu_i, \nu_j) - 1$, where $\nu_i = \partial_{ri}(\bar{D}_\ell(\lambda))$, and $(\bar{D}_\ell(\lambda))_{ii}$ is monic. Because $\nu_i = \kappa_i, i=1,\ldots,m$, where κ_i are the Kronecker indices of (A,B), thus $\partial_{ci}(\bar{D}_\ell(\lambda)) = \kappa_i$ and $\bar{D}_\ell(\lambda)$ is also column reduced. Therefore, $D_{rc}(\lambda)$, defined as $D_{rc}(\lambda) \triangleq \bar{D}_\ell(\lambda)D_{rhc}$, has $\partial_{ci}(D_{rc}(\lambda)) = \kappa_i$. From the assumption, D_{rhc} is an upper triangular matrix with diagonal elements all 1's. Thus, $D_{rc}(\lambda)$ is a column reduced canonical λ-matrix with column indices $\partial_{ci}(D_{rc}(\lambda)) = \kappa_i, i=1,\ldots,m$ and $D_{rc}(\lambda)$ is an appropriate closed-loop right characteristic λ-matrix. From Theorem 3.9 and $D_{rc}^{-1}(\lambda) = \bar{C}_J(\lambda I_n -\bar{A}_J)^{-1}\hat{\bar{B}}_J D_{rhc} + \hat{\bar{D}}_J D_{rhc}$, $\bar{C}_J$ contains all the preassigned right latent vectors, $\bar{q}_{ij}, j=0,1,\ldots,\ell_i-1, i=1,\ldots,k$, of $D_{rc}(\lambda)$. Therefore, the control law in Theorem 5.5 contains the preassigned right latent structure. ■

Corollary 5.4   With the linear state-feedback control law determined in Theorem 5.5, the right generalized eigenvector of the closed-loop system map $\bar{A} = A-BF$ can be determined by

$$\bar{q}_{aij} = T_c^{-1} \sum_{k=0}^{j} \frac{1}{k!} \psi_r^{(k)}(\bar{\lambda}_i)\bar{q}_{i(j-k)}, \quad 0 \leq j \leq \ell_i - 1 \qquad (5.14)$$

Proof:

Corollary 5.4 can be proved using Corollary 3.2. ■

Example 5.2

Let us consider the same open loop system described in Example 5.1. We shall construct the linear state feedback control laws using left and right latent structure assignments as follows:

(1) Left latent structure assignment

Let the desired latent roots and left latent vectors of the closed-loop right characteristic λ-matrix be

$$\bar{\lambda}_1 = -5 \; ; \qquad \bar{P}_{10} = [1 \; -2 \; 0]^T$$

$$\bar{\lambda}_2 = -4 \; ; \qquad \bar{P}_{20} = [3 \; -4 \; 0]^T$$

$$\bar{\lambda}_3 = -3 \; ; \qquad \bar{P}_{30} = [1 \; -1 \; 0]^T$$

$$\bar{\lambda}_4 = -2 \; ; \qquad \bar{P}_{40} = [1 \; 1 \; 0]^T$$

$$\bar{\lambda}_5 = -2 \; ; \qquad \bar{P}_{50} = [2 \; -3 \; 0]^T$$

From Theorem 5.4, we have

$$\bar{A}_J = \begin{bmatrix} -5 & 0 & 0 & 0 & 0 \\ 0 & -4 & 0 & 0 & 0 \\ 0 & 0 & -3 & 0 & 0 \\ 0 & 0 & 0 & -2 & 0 \\ 0 & 0 & 0 & 0 & -1 \end{bmatrix}$$

and

$$\bar{B}_J = \begin{bmatrix} 1 & -2 & 0 \\ 3 & -4 & 0 \\ 1 & -1 & 0 \\ 1 & 1 & 0 \\ 2 & -3 & 0 \end{bmatrix}$$

It can easily be checked that the Kronecker indices of $(\bar{A}_J, \bar{B}_J)$ are $\kappa_1 = 3$, $\kappa_2 = 2$, $\kappa_3 = 0$. Using the formulas given in Theorem 5.4, the feedback gain is found to be

$$F = - \begin{bmatrix} -35.4 & -19.2 & -46.6667 & -29.3333 & -17.6667 \\ 15.4 & -6.2 & -17.5000 & -9.5000 & -3.5000 \\ 0.0 & 0.0 & 0.0000 & 0.0000 & 0.0000 \end{bmatrix}$$

and the input gain is

$$G_F = \begin{bmatrix} 1 & -1 & -1 \\ 0 & 1 & -2 \\ 0 & 0 & 1 \end{bmatrix}$$

The linear state-feedback control law is

$$U(t) = -Fx(t) + G_F r(t)$$

The closed-loop canonical RMFD is

$$G_c(\lambda) = N_r(\lambda) D_{rc}^{-1}(\lambda)$$

where $D_{rc}(\lambda)$ is the closed-loop right characteristic λ-matrix:

$$D_{rc}(\lambda) = \begin{bmatrix} \lambda^3 + 7\lambda^2 + 22\lambda + 24 & 4\lambda + 8 & 0 \\ 6\lambda + 12 & \lambda^2 + 9\lambda + 14 & 0 \\ 0 & 0 & 1 \end{bmatrix}$$

It is straightforward to check the $D_{rc}(\lambda)$ has latent roots $\bar{\lambda}_i$ and left latent vectors $\bar{P}_{i0}$ for $1 \leq i \leq 5$, as desired.

(2) Right latent structure assignment

Let the desired latent roots and right latent vectors of the closed-loop right characteristic λ-matrix be

$$\bar{\lambda}_1 = -5 \; ; \qquad \bar{q}_{10} = [1 \; 0 \; 0]^T$$

$$\bar{\lambda}_2 = -4 \; ; \qquad \bar{q}_{20} = [1 \; 2 \; 0]^T$$

$$\bar{\lambda}_3 = -3 \; ; \qquad \bar{q}_{30} = [1 \; 0 \; 0]^T$$

$$\bar{\lambda}_4 = -2 \; ; \qquad \bar{q}_{40} = [1 \; 0 \; 0]^T$$

$$\bar{\lambda}_5 = -2 \; ; \qquad \bar{q}_{50} = [7 \; 6 \; 0]^T$$

From Theorem 5.5, we have

$$\bar{A}_J = \begin{bmatrix} -5 & 0 & 0 & 0 & 0 \\ 0 & -4 & 0 & 0 & 0 \\ 0 & 0 & -3 & 0 & 0 \\ 0 & 0 & 0 & -2 & 0 \\ 0 & 0 & 0 & 0 & -2 \end{bmatrix}$$

and

$$\bar{C}_J = \begin{bmatrix} 1 & 0 & 1 & 0 & 1 \\ 0 & 1 & 0 & 1 & -2 \\ 0 & 0 & 0 & 0 & 0 \end{bmatrix}$$

It is easy to check that the observability indices of $(\bar{A}_J, \bar{C}_J)$ are $\nu_1 = 3$, $\nu_2 = 2$, $\nu_3 = 0$ which are the Kronecker indices of the open loop system. Choosing the leading-column coefficient matrix of the closed-loop right characteristic λ-matrix as

$$D_{rhc} = \begin{bmatrix} 1 & 1 & 0 \\ 0 & 1 & 0 \\ 0 & 0 & 1 \end{bmatrix}$$

with formulas given in Theorem 5.5, the feedback gain is found to be

$$F = - \begin{bmatrix} -47.7 & -27.1 & -65.1667 & -44.8333 & -28.6667 \\ -10.6 & -3.8 & -11.0000 & -4.0000 & 0.0000 \\ 0.0 & 0 & 0.0000 & 0.0000 & 0.0000 \end{bmatrix}$$

and the input gain is

$$G_F = \begin{bmatrix} 1 & -2 & -1 \\ 0 & 1 & -2 \\ 0 & 0 & 1 \end{bmatrix}$$

The linear state-feedback control law is

$$u(t) = -FX(t) + G_F r(t)$$

The closed-loop canonical RMFD is found as

$$G_c(\lambda) = N_r(\lambda) D_{rc}^{-1}(\lambda)$$

where $D_{rc}(\lambda)$ is the closed-loop right characteristic λ-matrix:

$$D_{rc}(\lambda) = \begin{bmatrix} \lambda^3 + 10\lambda^2 + 31\lambda + 30 & 2\lambda^2 + 6.5\lambda + 9 & 0 \\ 0 & \lambda^2 + 6\lambda + 8 & 0 \\ 0 & 0 & 1 \end{bmatrix}$$

It is straightforward to check that $D_{rc}(\lambda)$ has latent roots $\bar{\lambda}_i$ and right latent vectors $\bar{q}_{i0}$ for $1 \leq i \leq 5$ as desired.

5.4 Divisor Assignment and Decoupling Design of MIMO Systems

In this section, we shall extend the latent structure assignment scheme discussed in Section 5.3 to the simpler state-feedback controller design method, namely, divisor assignment. Basically, the divisor assignment approach is to construct the feedback law by assigning the desired complete set of canonical left or right divisors for the closed-loop right characteristic λ-matrix. From Theorems 3.9 and 4.4, divisor assignment is equivalent to the latent structure assignment in Section 5.3. However, the notion of divisor assignment enables us to construct the state-feedback controllers for parallel forms (refer to Chapter VI for details), input decoupling, and input-output decoupling [69-71] of closed-loop MIMO systems.

<u>Theorem 5.6</u>   (Left Divisor Assignment)

Let $\{D_{rci}(\lambda), \ i=1,\ldots,k\}$ be a desired complete set of canonical left divisors of the closed-loop right characteristic λ-matrix. Assume that $(\bar{A}_{ci}, \bar{B}_{ci}, \bar{C}_{ci}, \bar{D}_{ci})$ is a minimal realization quadruple of $D_{rci}^{-1}(\lambda)$ for $i=1,\ldots,k$. Define

$$\bar{A}_c = \text{Block diag}[\bar{A}_{ci}, \ i=1,\ldots,k] \tag{5.15a}$$

and

$$\bar{B}_c = [\bar{B}_{c1}^T,\ldots,\bar{B}_{ck}^T] \tag{5.15b}$$

If the Kronecker indices of $(\bar{A}_c, \bar{B}_c)$ are the same as those of the open loop system, then the linear state-feedback control law in Eq. (5.2) can be constructed as

$$F = F_c T_c \tag{5.16a}$$
$$G_F = D_{rh} D_{rhc}^{-1} \tag{5.16b}$$

where F_c is determined by

$$F_c \psi_r(\lambda) = G_F D_{rc}(\lambda) - D_r(\lambda) \tag{5.16c}$$

Note that $D_r(\lambda)$ is the right characteristic λ-matrix of the open loop system described in Eq. (5.1); $D_{rc}(\lambda)$ is the right characteristic λ-matrix of $(\bar{A}_c, \bar{B}_c)$; D_{rh} and D_{rhc} are the leading-column coefficient matrices of $D_r(\lambda)$ and $D_{rc}(\lambda)$, respectively. $\psi_r(\lambda)$ is defined by Eq. (2.13b). T_c is the matrix which transforms the system map A of the open loop system to its controller canonical form A_c.

Proof:

Similar to the proof of Theorem 5.4.   ■

Theorem 5.7   (Right Divisor Assignment)

Let $\{D_{\ell c i}(\lambda), \ i=1,\ldots,k\}$ be a desired complete set of canonical right divisors of the closed-loop right characteristic λ-matrix. Assume that $(\bar{A}_{0i}, \bar{B}_{0i}, \bar{C}_{0i}, \bar{D}_{0i})$ is a minimal realization quadruple of $D_{\ell c i}^{-1}(\lambda)$ for $i=1,\ldots,k$. Define

$$\bar{A}_0 = \text{Block diag}[\bar{A}_{0i}, \ i=1,\ldots,k] \tag{5.17a}$$

and

$$\bar{C}_0 = [\bar{C}_{01},\ldots,\bar{C}_{0k}] \tag{5.17b}$$

If the observability indices of $(\bar{A}_0,\bar{C}_0)$ are the same as the Kronecker indices of the open loop system and the leading-row coefficient matrix of the left characteristic λ-matrix of. $(\bar{A}_0,\bar{C}_0)$ is I_m, then the linear state-feedback control law in Eq. (5.2) can be constructed as

$$F = F_c T_c \tag{5.18a}$$

$$G_F = D_{rh} D_{rhc}^{-1} \tag{5.18b}$$

where F_c is determined by

$$F_c \psi_r(\lambda) = G_F D_{rc}(\lambda) - D_r(\lambda) \tag{5.18c}$$

And, D_{rhc} is an upper triangular matrix with diagonal elements all 1's; $D_{rc}(\lambda) =$

D_{rhc} $\bar{D}_{\ell c}(\lambda)$ and $\bar{D}_{\ell c}(\lambda)$ is the left characteristic λ-matrix of $(\bar{A}_0, \bar{C}_0)$. $D_r(\lambda)$ is the right characteristic λ-matrix of the open loop system. $\psi_r(\lambda)$ is defined by Eq. (2.13b). T_c is the transformation to transform the system map A of the open loop system to its controller canonical form A_c.

Proof:

Similar to the proof of Theorem 5.5. ∎

In practical applications, we may choose the simpler structure for the complete set of canonical left/right divisors of the closed loop characteristic λ-matrix such that all the necessary conditions in Theorems 5.6 and 5.7 are satisfied. In Propositions 5.1-5.4, various ways to carry out the left/right divisor assignment are proposed. The information which is needed to perform the design of an MIMO system using the approaches in Propositions 5.1-5.4 is the set of Kronecker indices of the open loop system. In Proposition 5.5, we shall explore the input-output decoupling design problem, using the idea of divisor assignment.

Proposition 5.1   Let $D_{rci}(\lambda) = \text{diag}\{d_{ij}(\lambda), j=1,\ldots,m\}$ for $i=1,\ldots,k$, and $\sum_{i=1}^{k}$ $\deg[d_{ij}(\lambda)] = \kappa_j$, then $\{D_{rci}(\lambda), i=1,\ldots,k\}$ is a complete set of left divisors which can be assigned to construct $D_{rc}(\lambda)$. ∎

Since $D_{rc}(\lambda) = \text{diag}\{\prod_{i=1}^{k} d_{ij}(\lambda), j=1,\ldots,m\}$, the Kronecker indices of $D_{rc}(\lambda)$ will be the same as those of the open loop system. Applying Theorem 5.6 gives the desired feedback gains F and input gain G_F.

Similarly, for right divisor assignment, we have the following:

Proposition 5.2   Let $D_{\ell ci}(\lambda) = \text{diag}\{d_{ij}(\lambda), j=1,\ldots,m\}$ for $i=1,\ldots,k$, and $\sum_{i=1}^{k}$ $\deg[d_{ij}(\lambda)] = \kappa_j$, then $\{D_{\ell ci}(\lambda), i=1,\ldots,k\}$ is a complete set of right divisors which can be assigned to construct $D_{rc}(\lambda)$. ∎

Since $\bar{D}_{\ell c}(\lambda) = \text{diag}\{\prod\limits_{i=1}^{k} d_{ij}(\lambda), j=1,\ldots,m\}$, the observability indices of $\bar{D}_{\ell c}(\lambda)$ will be the same as the Kronecker indices of the open loop system. Selecting a D_{rhc} and applying Theorem 5.7 gives the desired feedback gain F and input gain G_F.

Proposition 5.3   Let

$$D_{rci}(\lambda) = \text{diag}[I_{i-1},d_i(\lambda),I_{m-i}],i=1,\ldots,m \qquad (5.19)$$

with $\deg[d_i(\lambda)] = \kappa_i$, then $\{D_{rci}(\lambda),i=1,\ldots,m\}$ is a complete set of left divisors which can be assigned to construct $D_{rc}(\lambda)$.   ∎

Proposition 5.4   Let

$$D_{\ell ci}(\lambda) = \text{diag}[I_{i-1},d_i(\lambda),I_{m-i}],i=1,\ldots,m \qquad (5.20)$$

with $\deg[d_i(\lambda)] = \kappa_i$, then $\{D_{\ell ci}(\lambda),i=1,\ldots,m\}$ is a complete set of right divisors which can be assigned to construct $D_{rc}(\lambda)$.   ∎

The divisors suggested in Proposition 5.3 result in

$$D_{rc}(\lambda) = \text{diag}\{d_i(\lambda),i=1,\ldots,m\}$$

and the closed-loop RMFD becomes

$$G_c(\lambda) = C[\lambda I_n-(A+BF)]^{-1}BG_F = \sum\limits_{i=1}^{m} N_{rci}(\lambda)D_{ri}^{-1}(\lambda)$$

Also, from Eq. (2.15c), we have $N_{rci}(\lambda) = [N_i(\lambda),0_{p\times 1},\ldots,0_{p\times 1}]$, where $N_i(\lambda)$ is a column polynomial vector. The closed-loop transfer function matrix becomes

$$y(t) = \sum_{i=1}^{m} \frac{N_i(\lambda)}{d_i(\lambda)} r_i(t) \tag{5.21}$$

The parallel decomposed system in Eq. (5.21) is an input decoupling system. The same reasoning can be applied for Proposition 5.4 with $D_{rhc} = I_m$. The technique for input decoupling can be extended to design an input-output decoupling system which is defined as

$$y_i(t) = \frac{n_i(\lambda)}{d_i(\lambda)} r_i(t) \text{ for } i=1,\ldots,m \text{ and } m=p \tag{5.22}$$

The necessary and sufficient conditions for input-output decoupling using linear state-feedback have been investigated by many authors [10,68-71]. Wolovich [10,69] derived the condition that the matrix $\lim\limits_{\lambda \to \infty} \xi_G(\lambda)G(\lambda) \triangleq K_G$ be nonsingular for $m=p$, where $\xi_G(\lambda)$ is the <u>interactor</u> of the open loop system, namely

$$\xi_G(\lambda) = \text{diag}\{\lambda^{f_i}, i=1,\ldots,m\} \quad \text{for} \quad f_i \geq 0$$

When K_G is nonsingular, the closed-loop characteristic λ-matrix can be factored as

$$D_{rc}(\lambda) = D_d(\lambda)N(\lambda) \tag{5.23a}$$

where $N(\lambda)$ is a right divisor of $N_r(\lambda)$, or

$$N_r(\lambda) = N_d(\lambda)N(\lambda) \tag{5.23b}$$

$$N_d(\lambda) = \text{diag}\{n_i(\lambda), i=1,\ldots,m\}; D_d(\lambda) = \text{diag}\{d_i(\lambda), i=1,\ldots,m\}$$

$$deg[d_i(\lambda)] = deg[n_i(\lambda)]+f_i$$

For practical applications, $N(\lambda)$ must be stable and invertible. Note that the closed-loop poles associated with $\prod_{i=1}^{m} d_i(\lambda)$ can be arbitrarily assigned, and that $n_i(\lambda)$ is the monic common divisor of the ith row component of $N_r(\lambda)$. The feedback gains for input-output decoupling can be determined as follows.

<u>Proposition 5.5</u>   Assume that $d_i(\lambda)$ are monic, then the feedback gains for the input-output decoupling are $F = F_c T_c, G_F = K_G^{-1}$, where F_c satisfies

$$F_c \psi_r(\lambda) = D_{rh}^{-1}[K_G^{-1}D_{rc}(\lambda)-D_r(\lambda)] \tag{5.24}$$

<u>Proof:</u>

Since

$$\lim_{\lambda\to\infty} \xi_G(\lambda)N_r(\lambda)D_{rc}^{-1}(\lambda) = K_G G_F$$

and

$$N_r(\lambda)D_{rc}^{-1}(\lambda) = diag\{n_i(\lambda)/d_i(\lambda), i=1,\ldots,m\}$$

where both $n_i(\lambda)$ and $d_i(\lambda)$ are monic, then $K_G G_F = I_m$ or $G_F = K_G^{-1}$. From Eq. (5.7), we obtain $G = D_{rh}D_{rch}^{-1}$ and $D_{rch} = K_G D_{rh}$. Thus, $F_c \psi_r(\lambda) = (K_G D_{rh})^{-1}D_{rc}(\lambda)-D_{rh}^{-1}D_r(\lambda) = D_{rh}^{-1}[K_G^{-1}D_{rc}(\lambda)-D_r(\lambda)]$. ∎

<u>Example 5.3</u>

Consider a 3-input, 2-output system

$$\lambda X(t) = AX(t)+Bu(t)$$
$$y(t) = CX(t)+Du(t)$$

where

$$A = \begin{bmatrix} -1 & -2 & 4 & -9 & 21 \\ 0 & -2 & 2 & -5 & 4 \\ 0 & 2 & -5 & 10 & -23 \\ 0 & 1 & -2 & 4 & -4 \\ 0 & 0 & 0 & 0 & 2 \end{bmatrix} ; \quad B = \begin{bmatrix} 4 & -3 & -2 \\ -1 & 1 & 1 \\ -1 & 3 & 5 \\ 2 & 1 & 4 \\ 1 & 0 & 1 \end{bmatrix}$$

$$C = \begin{bmatrix} 1 & -1 & 4 & -5 & 11 \\ 0 & 1 & -1 & 4 & -5 \end{bmatrix} ; \quad D = 0_{2\times 3}$$

The Kronecker indices are found to be $\kappa_1 = 3$, $\kappa_2 = 2$ and $\kappa_3 = 0$. Note that the input matrix B contains a dependent vector. From Eq. (2.11b) we obtain the similarity transformation matrix as

$$T_c = \begin{bmatrix} 0 & -0.11111 & 0.55556 & -1.55556 & 3.55556 \\ 0 & -0.22222 & 0.11111 & -0.11111 & 0.11111 \\ 0 & 0.55556 & -0.77778 & 1.77778 & -2.77778 \\ 1 & -2.88889 & 6.44444 & -13.4444 & 26.4444 \\ -1 & 3.22222 & -7.11111 & 16.1111 & -32.1111 \end{bmatrix}$$

A_c, B_c and C_c in the controller canonical form become

$$A_c = T_c A T_c^{-1} = \left[\begin{array}{ccc:cc} 0 & 1 & 0 & 0 & 0 \\ 0 & 0 & 1 & 0 & 0 \\ 2 & 3 & 0 & 0 & 0 \\ \hdashline 0 & 0 & 0 & 0 & 1 \\ -3 & -3 & 0 & -1 & -2 \end{array}\right] ; \quad B_c = T_c B = \left[\begin{array}{ccc} 0 & 0 & 0 \\ 0 & 0 & 0 \\ 1 & 0 & 1 \\ \hdashline 0 & 0 & 0 \\ 0 & 1 & 2 \end{array}\right]$$

$$C_c = C T_c^{-1} = \left[\begin{array}{ccc:cc} 2 & 0 & 2 & 4 & 3 \\ 4 & 4 & 3 & 2 & 2 \end{array}\right] ; \quad D_c = 0_{2\times 3}$$

From Eq. (2.14c) we obtain

$$\psi_r(\lambda) = \begin{bmatrix} 1 & \lambda & \lambda^2 & 0 & 0 \\ 0 & 0 & 0 & 1 & \lambda \\ 0 & 0 & 0 & 0 & 0 \end{bmatrix}^T ; \quad A_r = \begin{bmatrix} 2 & 3 & 0 & 0 & 0 \\ -3 & -3 & 0 & -1 & -2 \\ 0 & 0 & 0 & 0 & 0 \end{bmatrix} ;$$

$$D_{rh} = \begin{bmatrix} 1 & 0 & 1 \\ 0 & 1 & 2 \\ 0 & 0 & 1 \end{bmatrix}^{-1} = \begin{bmatrix} 1 & 0 & -1 \\ 0 & 1 & -2 \\ 0 & 0 & 1 \end{bmatrix}$$

$\psi_r(\lambda)$ plays an important role in determining the eigenvectors of the system matrix A_c and the latent vectors of the right characteristic λ-matrix of the system. The right characteristic λ-matrix of the system can be written as

$$D_r(\lambda) = D_{rh}[D_h(\lambda) - A_r\psi_r(\lambda)] = \begin{bmatrix} 1 & 0 & -1 \\ 0 & 1 & -2 \\ 0 & 0 & 1 \end{bmatrix} \begin{bmatrix} \lambda^3 - 3\lambda - 2 & 0 & 0 \\ 3\lambda + 3 & \lambda^2 + 2\lambda + 1 & 0 \\ 0 & 0 & 1 \end{bmatrix}$$

$$= \begin{bmatrix} \lambda^3 - 3\lambda - 2 & 0 & -1 \\ 3\lambda + 3 & \lambda^2 + 2\lambda + 1 & -2 \\ 0 & 0 & 1 \end{bmatrix}$$

$N_r(\lambda)$ in Eq. (2.15c) becomes

$$N_r(\lambda) = C_c\psi_r(\lambda) = \begin{bmatrix} 2\lambda^2 + 2 & 3\lambda + 4 & 0 \\ 3\lambda^2 + 4\lambda + 4 & 2\lambda + 2 & 0 \end{bmatrix}$$

The RMFD of the system can be described as $G(\lambda) = N_r(\lambda)D_r^{-1}(\lambda)$.

To demonstrate the input decoupling of a MIMO system, using the divisor assignment method, we assign the desired complete set of left divisors of the closed-loop right characteristic λ-matrix as:

$$D_{rc1}(\lambda) = \begin{bmatrix} \lambda^2 + 5\lambda + 4 & 0 & 0 \\ 0 & \lambda + 2 & 0 \\ 0 & 0 & 1 \end{bmatrix}, \quad D_{rc2}(\lambda) = \begin{bmatrix} \lambda + 2 & 0 & 0 \\ 0 & \lambda + 4 & 0 \\ 0 & 0 & 1 \end{bmatrix}$$

The minimal realization of the closed-loop characteristic λ-matrix in a Jordan form can easily be found from $D_{rc1}(\lambda)$ and $D_{rc2}(\lambda)$ as:

$$A_{JC} = \text{block diag}\left[\begin{pmatrix} -1 & 0 & 0 \\ 0 & -2 & 0 \\ 0 & 0 & -4 \end{pmatrix}, \begin{pmatrix} -2 & 0 \\ 0 & -4 \end{pmatrix} \right]; \quad B_{JC} = \begin{bmatrix} 1 & 0 & 1 & 1 & 0 \\ 0 & 1 & 0 & 0 & 1 \\ 0 & 0 & 0 & 0 & 0 \end{bmatrix}^T$$

$$C_{JC} = \begin{bmatrix} 0.33333 & 0.0 & 1.66667 & -0.5 & 0.0 \\ 0.0 & 0.5 & 0.0 & 0.0 & -0.5 \\ 0.0 & 0.0 & 0.0 & 0.0 & 0.0 \end{bmatrix}; \quad D_{JC} = \begin{bmatrix} 0 & 0 & 0 \\ 0 & 0 & 0 \\ 0 & 0 & 1 \end{bmatrix}$$

Thus, the right characteristic λ-matrix of the closed-loop system becomes

$$D_{rc}(\lambda) = \begin{bmatrix} \lambda^3+7\lambda^2+14\lambda+8 & 0 & 0 \\ 0 & \lambda^2+6\lambda+8 & 0 \\ 0 & 0 & 1 \end{bmatrix}$$

From Theorem 5.6, we have the state-feedback control law:

$$u(t) = FX(t) + Gr(t)$$

where

$$F = F_c T_c = \begin{bmatrix} 0.0 & 1.0 & -2.0 & 5.0 & -18.0 \\ -3.0 & 6.3333 & -14.66667 & 24.66667 & -45.66667 \\ 0.0 & 0.0 & 0.0 & 0.0 & 0.0 \end{bmatrix}$$

$$G_F = D_{rh}D_{rch}^{-1} = \begin{bmatrix} 1 & 0 & -1 \\ 0 & 1 & -2 \\ 0 & 0 & 1 \end{bmatrix}$$

The closed-loop RMFD of the designed system is described by

$$G_c(\lambda) = \begin{bmatrix} 2\lambda^2+2 & 3\lambda+4 & 0 \\ & & \\ 3\lambda^2+4\lambda+4 & 2\lambda+2 & 0 \end{bmatrix}\begin{bmatrix} \lambda^3+7\lambda^2+14\lambda+8 & 0 & 0 \\ 0 & \lambda^2+6\lambda+8 & 0 \\ 0 & 0 & 1 \end{bmatrix}^{-1}$$

The partial fraction expansion of the closed-loop RMFD yields (refer to Section 6.1)

$$G_c(\lambda) = \begin{bmatrix} 7\lambda+11 & -1 & 0 \\ 7\lambda+10 & -1 & 0 \end{bmatrix}\begin{bmatrix} \lambda^2+5\lambda+4 & 0 & 0 \\ 0 & \lambda+2 & 0 \\ 0 & 0 & 1 \end{bmatrix}^{-1} + \begin{bmatrix} -5 & 4 & 0 \\ -4 & 3 & 0 \end{bmatrix}\begin{bmatrix} \lambda+2 & 0 & 0 \\ 0 & \lambda+4 & 0 \\ 0 & 0 & 1 \end{bmatrix}^{-1}$$

Example 5.4

To demonstrate the input-output decoupling of a MIMO system, we consider the following 2-input, 2-output system:

$$\lambda X(t) = AX(t) + Bu(t)$$

$$y(t) = CS(t) + Du(t)$$

where

$$
A = \begin{bmatrix} -1 & -2 & 4 & -9 & 21 \\ 0 & -2 & 2 & -5 & 4 \\ 0 & 2 & -5 & 10 & -23 \\ 0 & 1 & -2 & 4 & -4 \\ 0 & 0 & 0 & 0 & 2 \end{bmatrix} \; ; \quad B = \begin{bmatrix} 4 & -3 \\ -1 & 1 \\ -1 & 3 \\ 2 & 1 \\ 1 & 0 \end{bmatrix}
$$

$$
C = \begin{bmatrix} 2 & -5\frac{7}{9} & 10\frac{8}{9} & -21\frac{8}{9} & 41\frac{8}{9} \\ -3 & 9 & -20 & 43 & -85 \end{bmatrix} \; ; \quad D = 0_{2 \times 2}
$$

The RMFD of the system can be computed as

$$
G(\lambda) = \begin{bmatrix} \lambda^2 + 3\lambda - 4 & -\lambda + 1 \\ 0 & \lambda - 2 \end{bmatrix} \begin{bmatrix} \lambda^3 - 3\lambda - 2 & 0 \\ 3\lambda + 3 & \lambda^2 + 2\lambda + 1 \end{bmatrix}^{-1}
$$

The interactor is

$$
\xi_G(\lambda) = \begin{bmatrix} \lambda & 0 \\ 0 & \lambda \end{bmatrix}
$$

$$
K_G = \lim_{\lambda \to \infty} \xi_G(\lambda) G(\lambda) = \begin{bmatrix} 1 & -1 \\ 0 & 1 \end{bmatrix}
$$

The numerator matrix $N_r(\lambda)$ can be decomposed as

$$
N_r(\lambda) = \begin{bmatrix} \lambda^3 + 3\lambda - 4 & -\lambda + 1 \\ 0 & \lambda - 2 \end{bmatrix} = N_d(\lambda) N(\lambda)
$$

where

$$
N_d(\lambda) = \begin{bmatrix} \lambda - 1 & 0 \\ 0 & \lambda - 2 \end{bmatrix} \; ; \quad N(\lambda) = \begin{bmatrix} \lambda + 4 & -1 \\ 0 & 1 \end{bmatrix}
$$

The number of poles which can be assigned is 4. Let the left divisor of the closed-loop right characteristic λ-matrix be

$$D_d(\lambda) = \begin{bmatrix} (\lambda+1)(\lambda+2) & 0 \\ 0 & (\lambda+2)(\lambda+4) \end{bmatrix}$$

Thus, the closed-loop right characteristic λ-matrix becomes

$$D_{rc}(\lambda) = D_d(\lambda)N(\lambda)$$

$$= \begin{bmatrix} (\lambda+1)(\lambda+2)(\lambda+4) & -(\lambda+1)(\lambda+2) \\ 0 & (\lambda+2)(\lambda+4) \end{bmatrix}$$

The feedback control law can be solved as

$$u(t) = G_F r(t) + FX(t)$$

where

$$G_F = K_G^{-1} = \begin{bmatrix} 1 & 1 \\ 0 & 1 \end{bmatrix}$$

$$F = F_c T_c = \begin{bmatrix} -3.0 & 8.66667 & -19.33333 & 37.33333 & -80.33333 \\ -3.0 & 6.33333 & -14.66667 & 24.66667 & -45.66667 \end{bmatrix}$$

The closed-loop RMFD of the designed system becomes

$$G_c(\lambda) = \begin{bmatrix} \dfrac{\lambda-1}{(\lambda+1)(\lambda+2)} & 0 \\ 0 & \dfrac{\lambda-2}{(\lambda+2)(\lambda+4)} \end{bmatrix}$$

showing that the designed system is input-output decoupled.

CHAPTER VI

STRUCTURAL DECOMPOSITION THEORIES AND APPLICATIONS IN
MULTIVARIABLE CONTROL SYSTEMS

The main purpose of this chapter is to investigate the interconnection structures of multivariable systems described by MFDs. The divisor and spectral factorization theorems developed in Chapter IV will be utilized as the primary tools for studying the properties of the basic interconnection structures, these being the parallel forms, the cascade forms, and the semi-cascade forms of multivariable systems.

For convenience in explaining these basic interconnection forms, we consider a class of m-input, m-output systems, which can be described by a proper, 3rd-degree, right matrix fraction description (RMFD) with a monic denominator λ-matrix [80]. The RMFD can be described by the following fundamental structures:

$$Y(\lambda) = G_r(\lambda)U(\lambda) \tag{6.1a}$$

$$G_r(\lambda) = N_r(\lambda)D_r^{-1}(\lambda) = (N_0\lambda^3+N_1\lambda^2+N_2\lambda+N_3)(I_m\lambda^3+D_1\lambda^2+D_2\lambda+D_3)^{-1} \tag{6.1b}$$

$$= N_0[(\lambda I_m+\hat{S}_1)(\lambda I_m+\hat{S}_2)(\lambda I_m+\hat{S}_3)][(\lambda I_m+\bar{S}_1)(\lambda I_m+\bar{S}_2)(\lambda I_m+\bar{S}_3)]^{-1} \tag{6.1c}$$

$$= (N_0\lambda^3+N_1\lambda^2+N_2\lambda+N_3)(\lambda I_m+\bar{S}_3)^{-1}(\lambda I_m+\bar{S}_2)^{-1}(\lambda I_m+\bar{S}_1)^{-1} \tag{6.1d}$$

$$= N_0+K_1(\lambda I_m+L_1)^{-1}+K_2(\lambda I_m+L_2)^{-1}+K_3(\lambda I_m+L_3)^{-1} \tag{6.1e}$$

$$= N_0[(\lambda I_m+Z_1)(\lambda I_m+P_1)^{-1}][(\lambda I_m+Z_2)(\lambda I_m+P_2)^{-1}][(\lambda I_m+Z_3)(\lambda I_m+P_3)^{-1}] \tag{6.1f}$$

where N_0, N_i and D_i, i=1,2,3 in Eq. (6.1b) are matrix coefficients; $\hat{S}_i$ and $\bar{S}_i$, i=1,2,3 in Eq. (6.1c) are the spectral factors of $N_r(\lambda)$ and $D_r(\lambda)$, respectively; $-L_i$, i=1,2,3 in Eq. (6.1e) are the complete set of left solvents [17-18,25-27] of $D_r(\lambda)$. The matrices K_i, i=1,2,3 in Eq. (6.1e) are the block residues [27-31] associated with L_i, i=1,2,3. P_i and Z_i, i=1,2,3 in Eq. (6.1f) are, in general, not the spectral factors or solvents of $N_r(\lambda)$ and $D_r(\lambda)$, respectively. Note that, for a single variable system, or a completely decoupled system, $\hat{S}_i = Z_i$ and $\bar{S}_i = L_i = P_i$ for i=1,2,3. The set of linear λ-matrices, $\{(\lambda I_m + L_i),$ i=1,2,3\}, in Eq. (6.1e) is the complete set of left divisors [31] of $D_r(\lambda)$.

The structure of Eq. (6.1d) is the semi-cascade form in which the denominator modes with associated latent vectors of $D_r(\lambda)$ are isolated from one another but not from the dynamic modes of the numerator. The semi-cascade form is a suitable structure for modal control of multivariable systems [30]. The parallel structure of Eq. (6.1e) is a block partial fraction form [26] of $G_r(\lambda)$, which is often used because of its satisfactory reliability and sensitivity properties in physical system realizations. The cascade structure of Eq. (6.1f) is valuable from many aspects. For example, such a structure provides advantages of simplicity in construction, sensitivity reduction, maintenance in physical system implementation and trouble shooting of the implementation hardware.

In this chapter, we shall investigate the interconnection structures of an m-input, p-output linear time-invariant system described by Eq. (2.1), repeated here for convenience:

$$\lambda X(t) = AX(t) + Bu(t) \tag{6.2a}$$
$$y(t) = CX(t) + Du(t) \tag{6.2b}$$

If the system in Eq. (6.2) is reachable, the canonical RMFD of the systems becomes (refer to Section 2.2)

$$G(\lambda) = N_r(\lambda)D_r^{-1}(\lambda) + D$$

where $N_r(\lambda)D_r^{-1}(\lambda)$ is a strictly proper rational λ-matrix, and $N_r(\lambda)$ and $D_r(\lambda)$ are right coprime. If the system in Eq. (6.2) is observable, the canonical LMFD of the system is given by (refer to Section 2.2)

$$G(\lambda) = D_{\ell}^{-1}(\lambda)N_{\ell}(\lambda)+D$$

where $D_{\ell}^{-1}(\lambda)N_{\ell}(\lambda)$ is a strictly proper rational λ-matrix, and $N_{\ell}(\lambda)$ and $D_{\ell}(\lambda)$ are left coprime. As pointed out in Section 2.1, for a general MIMO system, $D_{r}(\lambda)$ and $D_{\ell}(\lambda)$ may not be monic. Instead, they are canonical column-reduced and canonical row-reduced λ-matrices, respectively. The principal tools for the results developed in this chapter are the spectral decomposition/factorization theorems of nonsingular λ-matrices explored in Chapter IV via the geometric approaches [5,81-82].

This chapter is organized as follows. In Section 6.1, we study the parallel decomposition structure of MIMO systems and its applications to model reductions and multiport network synthesis. In Section 6.2, the semi-cascade realizations and their applications to modal control of MIMO systems are investigated. In Section 6.3, the minimal cascade realizations and their applications to multiport network synthesis are discussed.

6.1 Parallel Decomposition Theories and Their Applications to MIMO Systems and Circuits

In this section, a complete set of canonical left (right) divisors of the right (left) characteristic λ-matrix of a reachable (observable) MIMO system derived in Section 4 of Chapter IV is employed to study the parallel decomposed structures [26,27,31] of MIMO systems and to develop their applications to model reductions and network synthesis.

6.1.1 Parallel Decomposition Theories of MIMO Systems

As indicated in Theorem 4.13, if a MIMO system is reachable (observable), a complete set of canonical left (right) divisors of the right (left) characteristic λ-matrix exists.

<u>Definition 6.1</u>   Let a system (A,B,C,D) in Eq. (6.2) be reachable and decomposable into the forms in Eq. (4.30). Also, let $(\kappa_{i1}, \kappa_{i2}, \ldots, \kappa_{im})$ be the Kronecker indices of (A_i, B_i), and T_{ci} be the transformation matrix which transforms (A_i, B_i) into its controller canonical form (A_{ci}, B_{ci}). Define

$$A_{cp} \triangleq \text{block diag}[A_{ci}, i=1, \ldots, k]; \; A_{ci} \in C^{n_i \times n_i} \tag{6.3a}$$

$$B_{cp} \triangleq [B_{c1}^T, B_{c2}^T, \ldots, B_{ck}^T]^T; \; B_{ci} \in C^{n_i \times m} \tag{6.3b}$$

$$C_{cp} \triangleq [C_{c1}, C_{c2}, \ldots, C_{ck}]; \; C_{ci} \in C^{p \times n_i} \tag{6.3c}$$

$$D_{cp} \triangleq D \tag{6.3d}$$

and $A_{ci} = T_{ci} A_i T_{ci}^{-1}; \; B_{ci} = T_{ci} B_i; \; C_{ci} = C_i T_{ci}^{-1}$ for $i=1, \ldots, k$.   Then, $(A_{cp}, B_{cp}, C_{cp}, D_{cp})$ is called a <u>parallel controller canonical form</u> of the system (A,B,C,D). $\qquad\qquad\qquad\qquad\qquad\qquad\qquad\qquad\qquad\qquad\qquad\qquad\qquad$ $\square$

<u>Theorem 6.1</u> (The Generalized Right Partial Fraction Expansion Theorem)

Let $(A_{cp}, B_{cp}, C_{cp}, D_{cp})$ be a parallel controller canonical form of the system (A,B,C,D) in Eq. (6.2). Then, the right matrix fraction description (RMFD) can be decomposed as

$$G(\lambda) = N_r(\lambda) D_r^{-1}(\lambda) + D = \sum_{i=1}^{k} N_{ri}(\lambda) D_{ri}^{-1}(\lambda) + D_{cp} \tag{6.4a}$$

where

$$N_{ri}(\lambda) D_{ri}^{-1}(\lambda) \triangleq C_{ci}(\lambda I_{n_i} - A_{ci})^{-1} B_{ci}$$

$D_{ri}(\lambda)$ is the right characteristic λ-matrix of (A_{ci}, B_{ci}) and

$$N_{ri}(\lambda) = C_{ci}\psi_{ri}(\lambda) = C_i T_{ci}^{-1}\psi_{ri}(\lambda) \tag{6.4b}$$

where

$$\psi_{ri}(\lambda) \triangleq \begin{bmatrix} 1 & \lambda & \dots & \lambda^{\kappa_{i1}-1} & 0 & 0 & \dots & & \lambda^{\kappa_{i2}-1} & \dots & 0 & 0 & \dots & & 0 \\ 0 & 0 & \dots & \cdot & 1 & \lambda & \dots & \lambda^0 & & \dots & 0 & 0 & \dots & & 0 \\ \cdot & \cdot & \dots & \cdot & \cdot & \cdot & \dots & \cdot & & \dots & \cdot & \cdot & \dots & & \\ 0 & 0 & \dots & 0 & 0 & 0 & \dots & 0 & & \dots & 1 & \lambda & \dots & & \lambda^{\kappa_{im}^{\circ}-1} \end{bmatrix}^T$$

$$\tag{6.4c}$$

Proof:

From Eq. (6.3) we obtain

$$G(\lambda) \triangleq C(\lambda I_n - A)^{-1}B + D = C_{cp}(\lambda I_n - A_{cp})^{-1}B_{cp} + D_{cp}$$

$$= \sum_{i=1}^{k} C_{ci}(\lambda I_{n_i} - A_{ci})^{-1}B_{ci} + D_{cp} = \sum_{i=1}^{k} N_{ri}(\lambda)D_{ri}^{-1}(\lambda) + D_{cp} \qquad \blacksquare$$

Equation (6.4) shows that the RMFD can be described by partial fraction expansion forms. The interpretation of the system structure in Eq. (6.4) is that the original system has been decomposed into the parallel connections of subsystems described by the canonical RMFD $N_{ri}(\lambda)D_{ri}^{-1}(\lambda)$, i=1,...,k.

The dual results of the parallel decomposition for observable systems are described as follows.

Definition 6.2   If the system (A,B,C,D) in Eq. (6.2) is observable and decomposable into the forms of Eq. (4.30), let $(\nu_{i1}, \nu_{i2}, \dots, \nu_{ip})$ be the observability indices of (A_i, C_i) and T_{0i} be the transformation matrix which transforms (A_i, C_i) to its observer canonical form (A_{0i}, C_{0i}). Define

$$A_{0p} \stackrel{\Delta}{=} \text{block diag}[A_{0i}, i=1,\ldots,k]; \; A_{0i} \epsilon C^{n_i \times n_i} \tag{6.5a}$$

$$B_{0p} \stackrel{\Delta}{=} [B_{01}^T, B_{02}^T, \ldots, B_{0k}^T]^T; \; B_{0i} \epsilon C^{n_i \times m} \tag{6.5b}$$

$$C_{0p} \stackrel{\Delta}{=} [C_{01}, C_{02}, \ldots, C_{0k}]; \; C_{0i} \epsilon C^{p \times n_i} \tag{6.5c}$$

$$D_{0p} \stackrel{\Delta}{=} D$$

and $A_{0i} = T_{0i}^{-1} A_i T_{0i}$; $B_{0i} = T_{0i}^{-1} B_i$; $C_{0i} = C_i T_{0i}$ for $i=1,\ldots,k$.

Then, $(A_{0p}, B_{0p}, C_{0p}, D_{0p})$ is called a __parallel observer canonical form__ of the system (A,B,C,D). $\qquad\qquad\qquad\qquad\qquad\qquad\qquad\qquad\qquad\qquad\qquad\qquad$ $\square$

__Theorem 6.2__ (The Generalized Left Partial Fraction Expansion Theorem)

Let $(A_{0p}, B_{0p}, C_{0p}, D_{0p})$ be a parallel observer canonical form of the system (A,B,C,D) in Eq. (6.2). Then, the LMFD of the system can be decomposed as

$$G(\lambda) \stackrel{\Delta}{=} D_\ell^{-1}(\lambda) N_\ell(\lambda) + D = \sum_{i=1}^{k} D_{\ell i}^{-1}(\lambda) N_{\ell i}(\lambda) + D_{0p} \tag{6.6a}$$

where

$$D_{\ell i}^{-1}(\lambda) N_{\ell i}(\lambda) \stackrel{\Delta}{=} C_{0i}(\lambda I_{n_i} - A_{0i})^{-1} B_{0i}$$

$D_{\ell i}(\lambda)$ is the left characteristic λ-matrix of (A_{0i}, C_{0i}) and

$$N_{\ell i}(\lambda) \stackrel{\Delta}{=} \psi_{\ell i}(\lambda) B_{0i} = \psi_{\ell i}(\lambda) T_{0i}^{-1} B_i \tag{6.6b}$$

where

$$\psi_{\ell i}(\lambda) = \begin{bmatrix} 1 & \lambda & \cdots & \lambda^{\nu_{i1}-1} & 0 & 0 & \cdots & \lambda^{\nu_{i2}-1} & \cdots & 0 & 0 & \cdots & 0 \\ 0 & 0 & \cdots & 0 & 1 & \lambda & \cdots & \lambda^{\nu_{i2}-1} & \cdots & 0 & 0 & \cdots & 0 \\ \cdot & \cdot & \cdots & \cdot & \cdot & \cdot & \cdots & \cdot & \cdots & \cdot & \cdot & \cdots & \\ 0 & 0 & \cdots & 0 & 0 & 0 & \cdots & 0 & \cdots & 1 & \lambda & \cdots & \lambda^{\nu_{ip}-1} \end{bmatrix}$$

(6.6c)

Proof:

Similar to Theorem 6.1. ∎

As pointed out in Section 4.4, the matrix sign algorithm can be employed to compute the transformations for block diagonalization of the system map A and to find out the complete set of canonical left/right divisors for the right/left characteristic λ-matrix of the system in Eq. (6.2). Therefore, the matrix sign algorithms, together with the minimal nice selection algorithms (Section 2.3) for determining the Kronecker indices of each parallel decomposed subsystem, constitute effective numerical methods for performing the left/right matrix fraction expansions for left/right MFDs of MIMO systems.

Applications of the parallel decomposition for MIMO systems are illustrated in the following subsections.

6.1.2 Model Reduction of MIMO Systems

A low-degree model of a high-degree MFD is often desirable for computer simulations and control system designs. A model reduction method for a class of MIMO systems described by MFDs with monic characteristic λ-matrices has been proposed by Shieh and Tsay [27]. This method is extended in this subsection for analysis and design [83] of general MIMO systems.

The basic idea for reducing a high order MIMO system to a lower order model is to decompose the system into dominant and non-dominant subsystems, and to discard the non-dominant subsystem which has faster dynamic responses. We define the dominant and non-dominant modes as follows.

Definition 6.3   The non-dominant modes of a continuous-time system are defined as the modes with $\text{Re}(\lambda_i) < \gamma_1$, where γ_1 is a real number, while the <u>dominant modes</u> are the modes having $\text{Re}(\lambda_i) > \gamma_1$. If the eigenvalues distribution of the original system is unknown, the real number γ_1 can be chosen as $\gamma_1 = \text{trace}(A)/n$, which is the arithmetic mean of the eigenvalues of A. For a discrete-time system, the non-dominant modes are the modes with $|\lambda_i| < \gamma_2$, where γ_2 is a positive real number, while the dominant modes are the modes having $|\lambda_i| > \gamma_2$. If the eigenvalues distribution of the original system is unknown, the positive real number γ_2 can be chosen as $\gamma_2 = |\sqrt[n]{\det(A)}|$, which is the geometric mean of the eigenvalues of A. $\qquad\qquad\qquad\qquad\qquad\qquad\qquad\qquad\qquad\qquad\qquad\qquad\square$

Two methods, namely the frequency-domain reduction method and the time-domain reduction method, are to be derived for the model reduction of MIMO systems based on the idea mentioned above.

The use of Theorem 6.1 yields a frequency-domain model reduction method for a right MFD as follows.

<u>Theorem 6.3</u>   If an MIMO system can be decomposed into parallel connections of two subsystems as

$$G(\lambda) = N_{r1}(\lambda)D_{r1}^{-1}(\lambda) + N_{r2}(\lambda)D_{r2}^{-1}(\lambda) + D_{cp} \qquad (6.7a)$$

such that $D_{r1}(\lambda)$ contains latent roots of dominant modes and $D_{r2}(\lambda)$ contains non-dominant modes, then the reduced-degree model of $G(\lambda)$ can be determined by

$$G_A(\lambda) = N_{r1}(\lambda)D_{r1}^{-1}(\lambda) + \hat{D}_{cp} \simeq G(\lambda) \qquad (6.7b)$$

where

$$\hat{D}_{cp} = N_{r2}(0)D_{r2}^{-1}(0) + D_{cp} \quad \text{(for continuous-time systems)} \qquad (6.7c)$$

and

$$\hat{D}_{cp} = N_{r2}(1)D_{r2}^{-1}(1) + D_{cp} \quad \text{(for discrete-time systems)} \qquad (6.7d)$$

$\hat{D}_{cp}$ is an p×m constant matrix derived for matching the zero-th time-moment of impulse response of G(t), or the steady-state values of the unit-step responses of the original stable system.　■

In a similar manner, by utilizing Theorem 6.2, we obtain a frequency-domain model reduction method for a left MFD as follows.

__Theorem 6.4__　If an MIMO system can be decomposed into parallel connections of two subsystems as

$$G(\lambda) = D_{\ell 1}^{-1}(\lambda)N_{\ell 1}(\lambda)+D_{\ell 2}^{-1}(\lambda)N_{\ell 2}(\lambda)+D_{0p} \qquad (6.8a)$$

such that $D_{\ell 1}(\lambda)$ and $D_{\ell 2}(\lambda)$ contain latent roots of dominant modes and non-dominant modes, respectively, then the reduced-degree model of $G(\lambda)$ can be determined by

$$G_A(\lambda) = D_{\ell 1}^{-1}(\lambda)N_{\ell 1}(\lambda)+\hat{D}_{0p} \cong G(\lambda) \qquad (6.8b)$$

where

$$\hat{D}_{0p} = D_{\ell 2}^{-1}(0)N_{\ell 2}(0)+D_{0p} \text{ (for continuous-time systems)} \qquad (6.8c)$$

and

$$\hat{D}_{0p} = D_{\ell 2}^{-1}(1)N_{\ell 2}(1)+D_{0p} \text{ (for discrete-time systems)} \qquad (6.8d)$$
■

Since $D_{r1}(\lambda)$ in Eq. (6.7b) and $D_{\ell 1}(\lambda)$ in Eq. (6.8b) are obtained from the induced pair (A_{L1},B_{L1}) and the embedded pair (A_{R1},C_{R1}) of (A,B) and (A,C), respectively, the matrix sign algorithm in Theorem 4.21 can be employed to compute the frequency-domain reduction models in Theorems 6.3 and 6.4.

In some control system problems, state-space reduction models are needed. For instance, the suboptimal control design and the approximate observer design are in this category. For these purposes, the time-domain method for the model reduction of an _observable_ MIMO system can be stated as follows.

__Theorem 6.5__   Let (A,B,C,D) be an observable continuous-time system and define

$$S_1 \overset{\Delta}{=} \text{Ind}[\text{Sign}^+(A-\gamma_1 I_n)] \epsilon C^{n \times n_1}$$

$$S_2 \overset{\Delta}{=} \text{Ind}[\text{Sign}^-(A-\gamma_1 I_n)] \epsilon C^{n \times n_2} \quad , \quad n_1 + n_2 = n$$

$$A_{R1} \overset{\Delta}{=} S_1^+ A S_1 \epsilon C^{n_1 \times n_1} \quad , \quad S_1^+ = (S_1^* S_1)^{-1} S_1^*$$

$$B_{R1} \overset{\Delta}{=} \tilde{S}_1 B \epsilon C^{n_1 \times m} \quad , \quad \tilde{S}_1 \overset{\Delta}{=} [I_{n_1} \ 0_{n_1 \times n_2}][S_1 \ S_2]^{-1}$$

$$C_{R1} \overset{\Delta}{=} C S_1 \epsilon C^{p \times n_1}$$

$$\hat{D}_{0p} \overset{\Delta}{=} C(-A)^{-1}B + D - C_{R1}(-A_{R1})^{-1}B_{R1} \epsilon C^{p \times m}$$

The reduced-order model with dominant modes of the original system having a non-zero initial vector $X(0)$ becomes

$$\lambda z(t) = A_{R1} z(t) + B_{R1} u(t)$$

$$y_s(t) = C_{R1} z(t) + \hat{D}_{0p} u(t) \simeq y(t)$$

$$z(0) = [C_{R1}^* C_{R1}]^{-1} C_{R1}^* [C X(0) - \hat{D}_{0p} u(0)] \quad \text{if } p \geq n_1$$

or

$$z(0) = C_{R1}^* [C_{R1} C_{R1}^*]^{-1} [C X(0) - \hat{D}_{0p} u(0)] \quad \text{if } p \leq n_1$$

The corresponding canonical LMFD of $(A_{R1}, B_{R1}, C_{R1}, \hat{D}_{0p})$ is

$$G_A(\lambda) = D_{\ell 1}^{-1}(\lambda)N_{\ell 1}(\lambda)+\hat{D}_{0p}$$

Proof:

Theorem 6.5 can be proved from Theorems 6.4 and 4.16 and Corollary 4.8. ∎

An alternative time-domain method for the model reduction of a <u>reachable</u> MIMO system can be stated in a similar way:

<u>Theorem 6.6</u>   Let (A,B,C,D) be a reachable continuous-time system. Define

$$V_1 \triangleq \{\text{Ind}[(\text{Sign}^+(A-\gamma_1 I_n))^T]\}^T \epsilon C^{n_1 \times n}$$

$$V_2 \triangleq \{\text{Ind}[(\text{Sign}^-(A-\gamma_1 I_n))^T]\}^T \epsilon C^{n_2 \times n}, \quad n_1+n_2 = n$$

$$A_{L1} \triangleq V_1 A V_1^+ \epsilon C^{n_1 \times n_1} \quad , \quad V_1^+ = V_1^*(V_1 V_1^*)^{-1}$$

$$B_{L1} \triangleq V_1 B \epsilon C^{n_1 \times m}$$

$$C_{L1} \triangleq C\tilde{V}_1 \epsilon C^{p \times n_1} \quad , \quad \tilde{V}_1 \triangleq [V_1^T V_2^T]^{-T}[I_{n_1} \ 0_{n_1 \times n_2}]^T$$

$$\hat{D}_{cp} \triangleq C(-A)^{-1}B+D-C_{L1}(-A_{L1})^{-1}B_{L1} \epsilon C^{p \times m}$$

Then, the reduced-order model with dominant models of the original system having a non-zero initial vector $X(0)$ becomes

$$\lambda z(t) = A_{L1}z(t)+B_{L1}u(t)$$

$$y_s(t) = C_{L1}z(t)+\hat{D}_{cp}u(t) \simeq y(t)$$

$$z(0) = [C_{L1}^*C_{L1}]^{-1}C_{L1}^*[CX(0)-\hat{D}_{cp}u(0)] \text{ if } p \geq n_1$$

or

$$z(0) = C_{L1}^*[C_{L1}C_{L1}^*]^{-1}[CX(0)-\hat{D}_{cp}u(0)] \text{ if } p \leq n_1$$

The corresponding canonical RMFD of $(A_{L1},B_{L1},C_{L1},\hat{D}_{cp})$ is

$$G_A(\lambda) = N_{r1}(\lambda)D_{r1}^{-1}(\lambda)+\hat{D}_{cp} \qquad \blacksquare$$

Again, the matrix sign algorithm in Theorem 4.21 can be used to compute the time-domain reduction models in Theorems 6.5 and 6.6.

Example 6.1

Given a 3-input 2-output continuous-time system described by the state-space equations (6.2) with $X(0) = 0_{5\times1}$, where

$$A = \begin{bmatrix} 0.5 & -3.0 & 10.5 & -21.5 & 39.0 \\ 0.5 & -2.0 & -3.5 & 6.5 & -5.0 \\ -1.0 & 2.0 & -9.0 & 17.0 & -30.0 \\ -0.5 & 1.0 & -2.5 & 4.5 & -25.0 \\ 0.0 & 0.0 & 0.0 & 0.0 & -8.0 \end{bmatrix} ; \quad B = \begin{bmatrix} -2.0 & 1.0 & 0.0 \\ 1.0 & -2.0 & -3.0 \\ 1.0 & 1.0 & 3.0 \\ 2.0 & 1.0 & 4.0 \\ 1.0 & 0.0 & 1.0 \end{bmatrix} ;$$

$$C = \begin{bmatrix} 10.0 & -15.0 & 41.0 & -65.0 & 131.0 \\ 5.0 & -2.0 & 10.0 & -2.0 & 6.0 \end{bmatrix} ; \quad D = 0_{2\times3}$$

Find: (1) The corresponding canonical LMFD of this system.

(2) The block decomposition of the MIMO system into two subsystems: one containing eigenvalues with real parts greater than $\gamma_1 (= \text{trace}(A)/n)$ and the other containing eigenvalues with real parts less than γ_1.

(3) The left block partial fraction expansion and the reduced-degree model of the LMFD.

(4) The reduced-order model of the original system.

Solution:

Using the minimal nice selection algorithm in Section 2.3, we find that the Kronecker indices of this system are $\kappa_1 = 3$, $\kappa_2 = 2$ and $\kappa_3 = 0$. Since $\kappa_3 = 0$, there exists a dependent vector in the input matrix B. The corresponding canonical LMFD can be found as

$$G(\lambda) = D_\ell^{-1}(\lambda) N_\ell(\lambda)$$

where

$$D_\ell(\lambda) = \begin{bmatrix} \lambda^3 + 5.164\lambda^2 + 5.319\lambda + 3.475 & -0.208\lambda - 1.658 \\ -0.649\lambda^2 - 6.853\lambda - 4.366 & \lambda^2 + 8.836\lambda + 6.688 \end{bmatrix}$$

and

$$N_\ell(\lambda) = \begin{bmatrix} 7\lambda^2 + 22.149\lambda - 2.074 & 16\lambda^2 + 80.627\lambda + 55.24 & 39\lambda^2 + 183.403\lambda + 108.406 \\ -4.545\lambda - 54.377 & 6.611\lambda + 35.366 & 8.678\lambda + 16.354 \end{bmatrix}$$

We shall decompose the systems into the two subsystems described in (2) above, where eigenvalues with real parts greater than $\gamma_1 = \text{trace}(A)/n = -14/5 = -2.8$.

First, we have to compute $\text{Sign}^+(A - \gamma_1 I_5)$ and $\text{Sign}^-(A - \gamma_1 I_5)$, which can be expressed as

$$\text{Sign}^+(A + 2.8 I_5) = \frac{1}{2}\left[\text{Sign}(A + 2.8 I_5) + I_5\right]$$

$$\text{Sign}^-(A + 2.8 I_5) = I_5 - \text{Sign}^+(A + 2.8 I_5)$$

The computed matrix sign functions are:

$$\text{Sign}^+(A + 2.8 I_5) = \begin{bmatrix} 1 & 0 & 1 & -2 & 4 \\ 0 & 1 & -2 & 4 & -8 \\ 0 & 0 & 0 & 2 & -4 \\ 0 & 0 & 0 & 1 & -2 \\ 0 & 0 & 0 & 0 & 0 \end{bmatrix} ; \quad \text{Sign}^-(A + 2.8 I_5) = \begin{bmatrix} 0 & 0 & -1 & 2 & -4 \\ 0 & 0 & 2 & -4 & 8 \\ 0 & 0 & 1 & -2 & 4 \\ 0 & 0 & 0 & 0 & 2 \\ 0 & 0 & 0 & 0 & 1 \end{bmatrix}$$

From the ranks of $\text{Sign}^+(A+2.8I_5)$ and $\text{Sign}^-(A+2.8I_5)$, we have $n_1 = 3$ and $n_2 = 2$, respectively. Thus, the system map A can be block-decomposed into two submatrices by using the block modal matrix M_S. To find the block modal matrix M_S in Eq. (4.41b) for the block-diagonalization of A, we determine the canonical injection maps S_1 and S_2 in Eq. (4.40) using Algorithm 4.1 in Section 4.4. The obtained matrix M_S becomes

$$M_S = [S_1, S_2] = \begin{bmatrix} 1 & 0 & -2 & -1 & -4 \\ 0 & 1 & 4 & 2 & 8 \\ 0 & 0 & 2 & 1 & 4 \\ 0 & 0 & 1 & 0 & 2 \\ 0 & 0 & 0 & 0 & 1 \end{bmatrix}$$

where $S_1 = \text{Ind}[\text{Sign}^+(A+2.8I_5)]$ and $S_2 = \text{Ind}[\text{Sign}^-(A+2.8I_5)]$.

Let $\quad z(t) = M_S^{-1}x(t)$

where $\quad z(t) = [z_s^T(t), z_f^T(t)]^T$; $z(t) \epsilon C^{n \times 1}$; $z_s(t) \epsilon C^{n_1 \times 1}$; $z_f(t) \epsilon C^{n_2 \times 1}$.

Then, the block-decomposed system in z(t)-coordinates becomes

$$\lambda z(t) = A_D z(t) + B_D u(t)$$

$$y(t) = C_D z(t) + D_D u(t)$$

where

$$A_D = M_S^{-1} A M_S = \text{block diag}[A_{R1}, A_{R2}] \ ;$$

$$A_{R1} = S_1^+ A S_1 = \begin{bmatrix} -0.5 & -1.0 & -4.5 \\ 2.5 & -6.0 & -27.5 \\ -0.5 & 1.0 & 4.5 \end{bmatrix} \ ; \quad A_{R2} = S_2^+ A S_2 = \begin{bmatrix} -4.0 & 20.0 \\ 0.0 & -8.0 \end{bmatrix}$$

$$B_D = M_S^{-1} B = [B_{R1}^T, B_{R2}^T]^T \ ;$$

$$B_{R1} = \begin{bmatrix} -1.0 & 2.0 & 3.0 \\ -1.0 & -4.0 & -9.0 \\ 0.0 & 1.0 & 2.0 \end{bmatrix} ; \quad B_{R2} = \begin{bmatrix} -3.0 & -1.0 & -5.0 \\ 1.0 & 0.0 & 1.0 \end{bmatrix}$$

$$C_D = CM_S = [C_{R1}, C_{R2}] ;$$

$$C_{R1} = CS_1 = \begin{bmatrix} 10.0 & -15.0 & -63.0 \\ 5.0 & -2.0 & 0.0 \end{bmatrix} ; \quad C_{R2} = CS_3 = \begin{bmatrix} 1.0 & 5.0 \\ 1.0 & 6.0 \end{bmatrix}$$

$$D_D = D = 0_{2 \times 3}$$

From Theorem 6.2, we obtain the left block partial fraction expansion of $G(\lambda)$ as

$$G(\lambda) = D_{\ell 1}^{-1}(\lambda) N_{\ell 1}(\lambda) + D_{\ell 2}^{-1}(\lambda) N_{\ell 2}(\lambda)$$

where

$$D_{\ell 1}(\lambda) = \begin{bmatrix} \lambda^2 + 1.154\lambda + 0.903 & -0.469 \\ -0.810\lambda - 0.565 & \lambda + 0.846 \end{bmatrix}; \quad N_{\ell 1}(\lambda) = \begin{bmatrix} 5\lambda - 0.232 & 17\lambda + 13.611 & 39\lambda + 27 \\ -7.052 & 4.225 & 1.398 \end{bmatrix} ;$$

$$D_{\ell 2}(\lambda) = \begin{bmatrix} \lambda + 4 & 0 \\ -4 & \lambda + 8 \end{bmatrix} ; \quad N_{\ell 2}(\lambda) = \begin{bmatrix} 2 & -1 & 0 \\ 3 & -1 & 1 \end{bmatrix}$$

Note that $\{D_{\ell 1}(\lambda), D_{\ell 2}(\lambda)\}$ is a complete set of canonical right divisors of $D_\ell(\lambda)$.

Applying Theorem 6.4, the reduced-degree model obtained by dropping the subsystem $D_{\ell 2}^{-1}(\lambda) N_{\ell 2}(\lambda)$, which has eigenvalues less than -2.8, becomes

$$G_A(\lambda) = D_{\ell 1}^{-1}(\lambda) N_{\ell 1}(\lambda) + \hat{D}_{0p} \simeq G(\lambda)$$

where

$$\hat{D}_{0p} = D_{\ell 2}^{-1}(0) N_{\ell 2}(0) = \begin{bmatrix} -0.5 & 0.25 & 0.0 \\ -0.625 & 0.25 & -0.125 \end{bmatrix}$$

The reduced-order model obtained from Theorem 6.5 becomes

$$\lambda z_s(t) = A_{R1} z_s(t) + B_{R1} u(t)$$

$$y_x(t) = C_{R1} z_s(t) + \hat{D}_{0p} u(t) \simeq y(t)$$

$$z_s(0) = (C_{R1}^* C_{R1})^{-1} C_{R1}^* [C_D z(0) - \hat{D}_{0p} u(0)]$$

$$= \begin{bmatrix} 1.164 & -0.465 & 2.327 & -0.465 & 1.396 \\ 0.643 & -0.257 & 1.285 & -0.257 & 0.771 \\ -0.539 & 0.903 & -2.390 & 4.028 & -8.084 \end{bmatrix} z(0)$$

$$+ \begin{bmatrix} -0.145 & 0.058 & -0.029 \\ -0.080 & 0.032 & -0.016 \\ -0.021 & -0.011 & -0.002 \end{bmatrix} u(0)$$

The time-response curves of the original system and the reduced-degree model and reduced-order models obtained above for the unit step inputs of $u(t) = [1,0,0]^T$, $u(t) = [0,1,0]^T$ and $u(t) = [0,0,1]^T$ are compared in Figs. 6.1-6.6. The approximation is quite satisfactory. It is interesting to note that the eigenvalues of the original system are $\lambda_{1,2} = -0.5 \pm j0.5$, $\lambda_3 = -1.0$, $\lambda_4 = -4.0$ and $\lambda_5 = -8$. The reduced-degree model and the reduced-order model obtained by using Theorems 6.11 and 6.5, respectively, contain eigenvalues λ_1, λ_2 and λ_3.

6.1.3 Multiport Network Synthesis

If $G(\lambda)$ is the transfer function matrix of a multiport network to be realized, then Eqs. (6.4a) and (6.6a) are the parallel decompositions of $G(\lambda)$. Each subsystem $N_{ri}(\lambda) D_{ri}^{-1}(\lambda)$ can be realized using active elements or R-C ladder networks with summers [84,85]. Thus, a desirable network can be constructed by connecting all subnetworks together with the direct matrix gain D_{cp}. The proposed approach can be viewed as the generalized Foster realization [86-88] for multiport network synthesis.

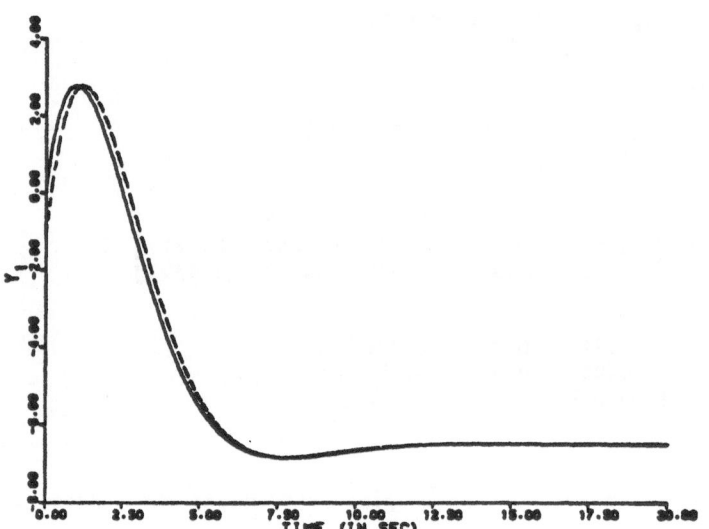

FIG. 6.1 RESPONSE OF $Y_1(t)$ TO INPUTS : $U_1=1.U_2=0.U_3=0$

———————— ORIGINAL SYSTEM

------ REDUCED MODEL

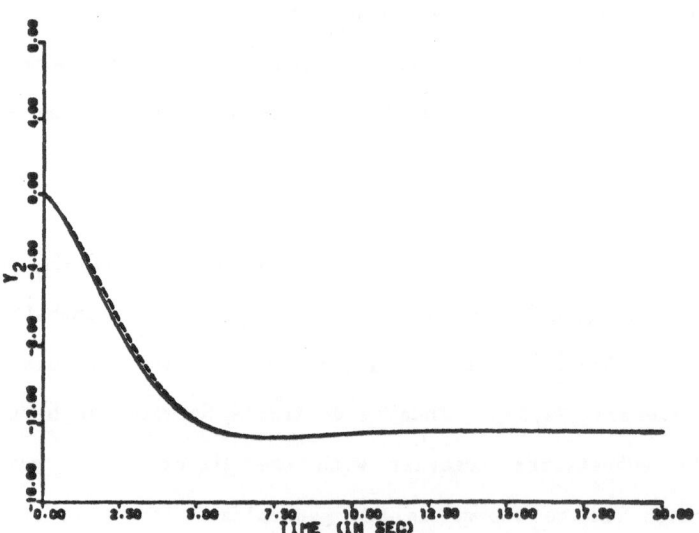

FIG. 6.2 RESPONSE OF $Y_2(t)$ TO INPUTS : $U_1=1.U_2=0.U_3=0$

———————— ORIGINAL SYSTEM

------ REDUCED MODEL

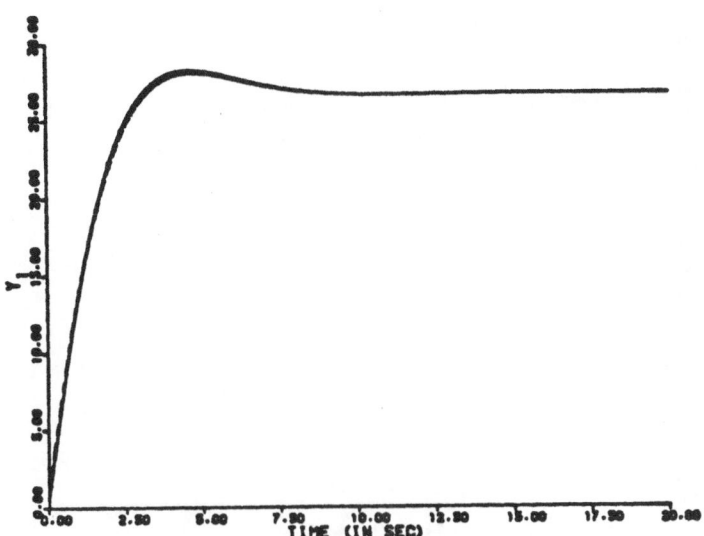

FIG. 6.3 RESPONSE OF $Y_1(t)$ TO INPUTS : $U_1=0.U_2=1.U_3=0$

——————— ORIGINAL SYSTEM

- - - - - - REDUCED MODEL

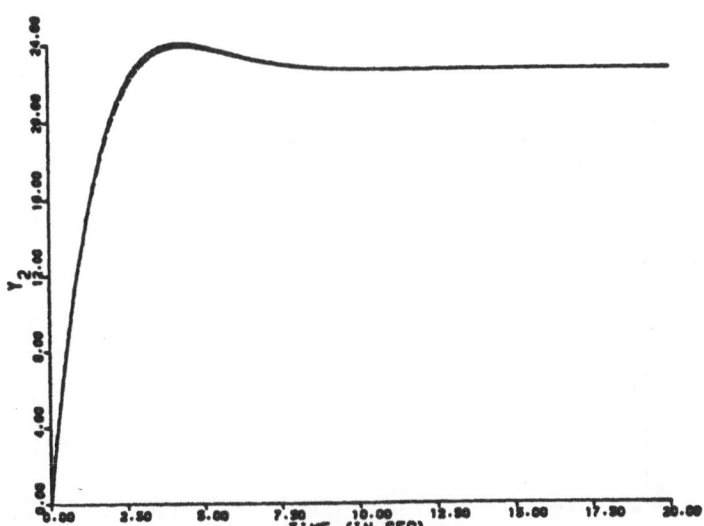

FIG. 6.4 RESPONSE OF $Y_2(t)$ TO INPUTS : $U_1=0.U_2=1.U_3=0$

——————— ORIGINAL SYSTEM

- - - - - - REDUCED MODEL

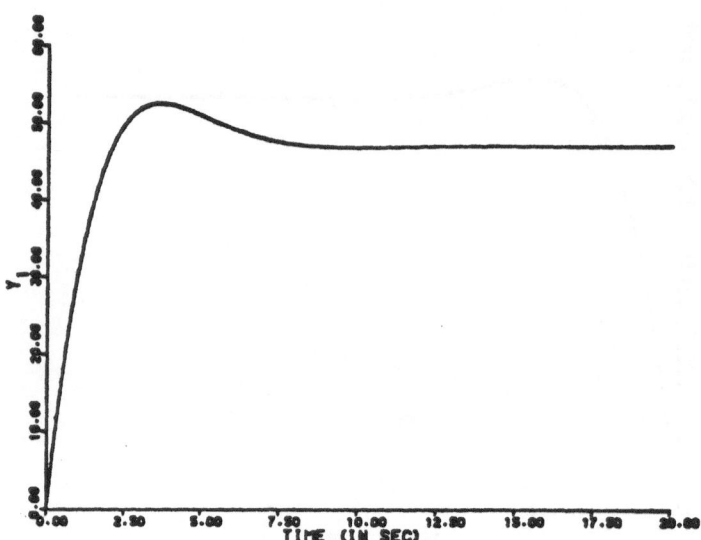

FIG. 6.5 RESPONSE OF $Y_1(t)$ TO INPUTS : $U_1=0.U_2=0.U_3=1$

———— ORIGINAL SYSTEM

------ REDUCED MODEL

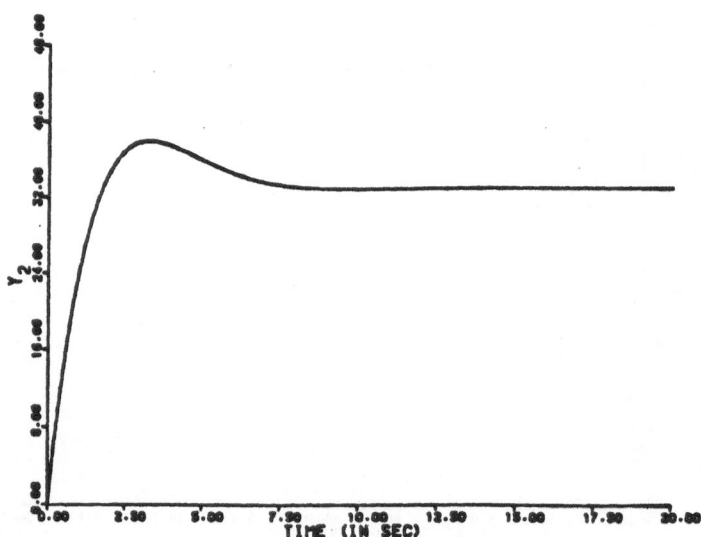

FIG. 6.6 RESPONSE OF $Y_2(t)$ TO INPUTS : $U_1=0.U_2=0.U_3=1$

———— ORIGINAL SYSTEM

------ REDUCED MODEL

6.2   Semi-Cascade Decomposition Theories and Their Applications to Modal
Controls of MIMO Systems

In this section, we investigate the realization of MFDs via spectral
factorizations of the characteristic λ-matrices of MIMO systems, and develop
their applications to modal control of MIMO systems [30,89,90].

6.2.1 Semi-Cascade Decomposition Theories

If the MIMO system (A,B,C,D) of Eq. (6.2) is reachable and A,B,C can be
decomposed into the forms in Eq. (4.28), then from Theorem 4.12, we obtain the
RMFD as

$$G(\lambda) = N_r(\lambda)[D_{r2}(\lambda)R(\lambda)]^{-1}+D = N_r(\lambda)R^{-1}(\lambda)D_{r2}^{-1}(\lambda)+D \qquad (6.9)$$

where $D_{r2}(\lambda)$ is the right characteristic λ-matrix of (A_2,B_2) and $R(\lambda)$ is a right
divisor of $D_r(\lambda)$.

Lemma 6.1   A reachable MIMO system of dimension greater than 1 can always be
decomposed into a semi-cascade realization form as follows.

$$\lambda X(t) = AX(t) + Bu(t) \qquad (6.10a)$$
$$y(t) = CX(t) + Du(t) \qquad (6.10b)$$

where

$$A = \begin{bmatrix} A_1 & A_{12} \\ 0 & A_2 \end{bmatrix} ; \quad A_i \in C^{n_i \times n_i}; \ i=1,2 \qquad (6.10c)$$

$$B = [B_1^T, B_2^T]^T; \ B_i \in C^{n_i \times m}; \ i=1,2 \qquad (6.10d)$$

$$C = [C_1, C_2] \; ; \; C_i \epsilon C^{p \times n_i}; \; i=1,2 \qquad\qquad (6.10e)$$

$$D = 0_{p \times m} \qquad\qquad (6.10f)$$

The corresponding semi-cascade RMFD is given by

$$G(\lambda) = N_r(\lambda) R^{-1}(\lambda) D_{r2}^{-1}(\lambda) \qquad\qquad (6.11a)$$

where $D_{r2}(\lambda)$ is the right characteristic λ-matrix of (A_2, B_2), and it has $(A_2, B_2, C_{r2}, D_{r2})$ as its associated minimal realization, where C_{r2} and D_{r2} are defined in Lemma 2.1. Also

$$N_r(\lambda) = C T_c^{-1} \psi_r(\lambda) \qquad\qquad (6.11b)$$

$$R(\lambda) = C_{r2} V T_c^{-1} \psi_r(\lambda) + D_{r2} D_r(\lambda) \qquad\qquad (6.11c)$$

where $D_r(\lambda)$ is the right characteristic λ-matrix of (A,B) and V is the canonical projection $X \rightarrow X/S$ in which (A_2, B_2) is induced, where S is an A-invariant subspace of X which is the state-space.

Proof:

Lemma 6.1 can be proved by directly using Theorem 4.1 and Theorem 4.12. ∎

Lemma 6.1 reveals the state-space structure and the RMFD representation for the semi-cascade realization of a reachable MIMO system. From Eqs. (6.10) and (6.11), the semi-cascade realization of a reachable system is, in fact, the result of spectral factorization of the right characteristic λ-matrix of the system.

The dual results for the semi-cascade realizations of observable systems can be stated as follows.

<u>Lemma 6.2</u>   An observable MIMO system in Eq. (6.2) can always be decomposed into a semi-cascade realization form as shown in Eq. (6.10) and the corresponding semi-cascade LMFD is given by

$$G(\lambda) = D_{\ell 1}^{-1}(\lambda)L^{-1}(\lambda)N_{\ell}(\lambda) \qquad\qquad (6.12a)$$

where $D_{\ell 1}(\lambda)$ is the left characteristic λ-matrix of (A_1, C_1) and it has $(A_1, B_{\ell 1}, C_1, D_{\ell 1})$ as its minimal realization. Also

$$N_{\ell}(\lambda) = \psi_{\ell}(\lambda)T_0^{-1}B \qquad\qquad (6.12b)$$

$$L(\lambda) = \psi_{\ell}(\lambda)T_0^{-1}SB_{\ell 1} + D_{\ell}(\lambda)D_{\ell 1} \qquad\qquad (6.12c)$$

where $D_{\ell}(\lambda)$ is the left characteristic λ-matrix of (A,C), S is the canonical injection map from $S{\to}X$, and (A_1, C_1) is the embedded map of (A,C) in S, where X and S are defined in Lemma 6.1.

<u>Proof</u>:

   Similar to Lemma 6.1.   ∎

6.2.2 Modal Control of MIMO Systems

   To perform modal control of MIMO systems by assigning the divisors of the characteristic λ-matrices, which contain specific latent roots and associated latent vectors of the λ-matrices, we study the MIMO system (A,B,C,D) in Eq. (6.2) which can be transformed into the following Jordan canonical form, using the modal matrix M of A:

$$\lambda X_J(t) = A_J X_J(t) + B_J u(t) \tag{6.13a}$$

$$y(t) = C_J X_J(t) \tag{6.13b}$$

where

$$X_J(t) = M^{-1} X(t) \tag{6.13c}$$

$$A_J = M^{-1} A M = \text{block diag}[A_{Ji}, i=1,\ldots,k]; A_{Ji} \epsilon C^{n_i \times n_i} \tag{6.13d}$$

$$B_J = M^{-1} B = [B_{J1}^T, B_{J2}^T, \ldots, B_{Jk}^T]^T; B_{Ji} \epsilon C^{n_i \times m} \tag{6.13e}$$

$$C_J = C M = [C_{J1}, C_{J2}, \ldots, C_{Jk}]; C_{Ji} \epsilon C^{p \times n_i} \tag{6.13f}$$

Each A_{Ji} is a full Jordan block with eigenvalue λ_i, i.e.

$$A_{Ji} = \begin{bmatrix} \lambda_i & 1 & \cdot & \cdot & \cdot \\ \cdot & \lambda_i & 1 & \cdot & \cdot \\ \cdot & \cdot & \cdot & \cdot & \cdot \\ \cdot & \cdot & \cdot & \cdot & \lambda_i \end{bmatrix} \epsilon C^{n_i \times n_i} \tag{6.13g}$$

The generalized left eigenvectors, $P_{Ji} \epsilon C^{n \times n_i}$, associated with A_{Ji} can be determined by partitioning the modal matrix M such that the following matrix equations are satisfied:

$$A p_{Ji} = p_{Ji} A_{Ji}, \quad i=1,2,\ldots,k \tag{6.13h}$$

The goal of modal control is to design a linear state-feedback control law for the system in Eq. (6.2), or the equivalent system in Eqs. (6.10) and (6.13), such that some undesirable eigenvalues and associated eigenvectors of the system can be replaced by desirable ones. Since the system map of Eq. (6.13) can be viewed as a special case of the block triangularized system map in Eq. (6.10), we consider the more general block triangularized system in Eq. (6.13) as

$$\lambda X_D(t) = A_D X_D(t) + B_D u(t) \qquad (6.14a)$$

$$y(t) = C_D X_D(t) + D_D u(t) \qquad (6.14b)$$

$$X_D(t) = T_D^{-1} X(t) \qquad (6.14c)$$

where

$$A_D = \begin{bmatrix} A_{D1} & A_{D12} \\ 0 & A_{D2} \end{bmatrix}; \quad B_D = \begin{bmatrix} B_{D1} \\ B_{D2} \end{bmatrix}; \quad X_D(t) = \begin{bmatrix} X_{D1}(t) \\ X_{D2}(t) \end{bmatrix};$$

$$C_D = [C_{D1}, C_{D2}]; \quad D_D = 0$$

A_{D1} is the collection of the unimportant system modes <u>not</u> to be controlled, and A_{D2} contains all dynamic modes to be controlled. $X_{Di}(t)$, B_{Di} and C_{Di} are matrices of appropriate dimensions. Let the control law be

$$u(t) = -F_D X_D(t) + G_F r(t) = -F_D T_D^{-1} X(t) + G_F r(t) \qquad (6.15a)$$

where $F_D \overset{\Delta}{=} [0, F_{D2}]$. Then, the closed-loop system becomes

$$\lambda X_D(t) = \bar{A}_D X_D(t) + \bar{B}_D u(t) \qquad (6.15b)$$

$$y(t) = C_D X_D(t) \qquad (6.15c)$$

where

$$\bar{A}_D = A_D - B_D F_D = \begin{bmatrix} A_{D1} & A_{D12} - B_{D1}F_{D2} \\ 0 & A_{D2} - B_{D2}F_{D2} \end{bmatrix} ; \quad \bar{B}_D = \begin{bmatrix} B_{D1}G_F \\ B_{D2}G_F \end{bmatrix} = B_D G_F$$

The modal control design problem is to find F_{D2} such that $A_{D2} - B_{D2}F_{D2}$ has the prespecified eigenvalues and eigenvectors. From the structures of $\bar{A}_D$ and $\bar{B}_D$, we observe that the modal control of the entire system is equivalent to the eigenstructure assignment of the partitioned subsystem (A_{D2}, B_{D2}). Using Theorem 5.1, we obtain the control law for the sub-system (A_{D2}, B_{D2}) as follows.

Lemma 6.3   Let $D_{r2}(\lambda)$ be the right characteristic λ-matrix of (A_{D2}, B_{D2}) which contains the undesirable modes to be controlled. Assume that the desirable closed-loop right characteristic λ-matrix with prespecified modes is $D_{rc2}(\lambda)$. Then, the feedback gain F_{D2} and input gain G_F in Eq. (6.15) become $F_{D2} = F_{c2}T_{c2}$ and $G_F = D_{rh2}D_{rhc2}^{-1}$, respectively. F_{c2} can be determined from the following matrix equation:

$$F_{c2}\psi_{r2}(\lambda) = G_F D_{rc2}(\lambda) - D_{r2}(\lambda)$$

T_{c2} is the transformation matrix which transforms (A_{D2}, B_{D2}) into its canonical controller form, and D_{rh2} and D_{rhc2} are the leading-column matrices of $D_{r2}(\lambda)$ and $D_{rc2}(\lambda)$, respectively. ∎

From the structure of the closed-loop state equations in Eq. (6.15) and Lemma 6.1, we observe that $D_{rc2}(\lambda)$ is the canonical left divisor of the closed-loop characteristic λ-matrix, $D_{rc}(\lambda)$. Therefore, the latent roots and left generalized latent vectors of $D_{rc2}(\lambda)$ are also part of the latent roots and left generalized latent vectors of $D_{rc}(\lambda)$. As a result, we can determine the modal control law via latent roots and associated latent vectors assignment as follows.

Theorem 6.7   (Modal Control Law with Latent Structure Assignment)

Assume that the desired closed-loop latent roots and left latent vectors corresponding to the modes to be controlled are $(\bar{\lambda}_{2i}, i=1,\ldots,k)$ and $(\bar{p}_{2ij}, j=0,1,\ldots,\ell_i-1, i=1,\ldots,k)$, respectively. Define

$$\bar{A}_{J2} = \text{block diag}[\bar{J}_{2i}, i=1,\ldots,k] \in C^{n_2 \times n_2}; n_2 = \sum_{i=1}^{k} \ell_i \qquad (6.16a)$$

$$\bar{B}_{J2} = [\bar{B}_{J21}^T, \ldots, \bar{B}_{J2k}^T]^T \in C^{n_2 \times m} \qquad (6.16b)$$

where $\bar{J}_{2i}$ and $\bar{B}_{J2i}$ are defined in Theorem 5.4.

Assume that $(\bar{A}_{J2}, \bar{B}_{J2})$ is a reachable pair and that the Kronecker indices of $(\bar{A}_{J2}, \bar{B}_{J2})$ are equal to those of (A_{D2}, B_{D2}) in Eq. (6.14). Then, the feedback gain F_D and the input gain G_F of the modal control law in Eq. (6.15a) can be determined by $F_D = [0, F_{D2}]$, where $F_{D2} = F_{c2}T_{c2}$ and $G_F = D_{rh2}D_{rhc2}^{-1}$. F_{c2} can be determined from the following matrix equation:

$$F_{c2}\psi_{r2}(\lambda) = G_F D_{rc2}(\lambda) - D_{r2}(\lambda) \qquad (6.16c)$$

where $D_{r2}(\lambda)$, $\psi_{r2}(\lambda)$ are the same as those defined in Lemma 6.3, and $D_{rc2}(\lambda)$ is the right characteristic λ-matrix of $(\bar{A}_{J2}, \bar{B}_{J2})$.

Proof:

Theorem 6.7 can be proved from Theorem 5.4 and Lemma 6.3.   ■

From Theorems 5.2 and 6.3, the closed-loop RMFD of the system in Eq. (6.10) with $D = 0_{p \times m}$ and the modal control law in Eq. (6.15a) can be represented as

$$G_c(\lambda) = C(\lambda I_n - \bar{A})^{-1}BG_F = N_r(\lambda)\hat{D}_{rc}^{-1}(\lambda) \qquad (6.17a)$$

where

$$\bar{A} = A - BF_D T_D^{-1} \tag{6.17b}$$

$$\hat{D}_{rc}(\lambda) = G_F^{-1}[F_D T_D^{-1} T_c^{-1} \psi_r(\lambda) + D_r(\lambda)] \tag{6.17c}$$

and $N_r(\lambda) D_r^{-1}(\lambda) = G(\lambda)$ is the canonical RMFD of the open loop system in Eq. (6.10), T_c is the transformation matrix which transforms (A,B) into its controller canonical form. The properties of $\hat{D}_{rc}(\lambda)$ can be stated as follows.

<u>Theorem 6.8</u> $\hat{D}_{rc}(\lambda)$ defined in Eq. (6.17c) is column reduced. If the open loop right characteristic λ-matrix of the system in Eq. (6.14) is factored as

$$D_r(\lambda) = R(\lambda) D_{r2}(\lambda) \tag{6.18}$$

where $D_{r2}(\lambda)$ is the right characteristic λ-matrix of (A_{D2}, B_{D2}), then $\hat{D}_{rc}(\lambda)$ can be factored as

$$\hat{D}_{rc}(\lambda) = R(\lambda) D_{rc2}(\lambda) \tag{6.19}$$

where $D_{rc2}(\lambda)$ is the right characteristic λ-matrix of $(A_{D2} - B_{D2} F_{D2}, B_{D2} G_F)$.

<u>Proof:</u>

From Eq. (6.17c), the leading-column matrix, $G_F^{-1} D_{rh}$, is nonsingular because D_{rh}, the leading column matrix of $D_r(\lambda)$, is nonsingular, and therefore $\hat{D}_{rc}(\lambda)$ is column reduced.

If $D_r(\lambda)$ is factored as in Eq. (6.18), then from Theorem 4.1, Lemma 6.1 and Eq. (6.14), we obtain

$$D_{r2}(\lambda) = \bar{C}_{r2}(\lambda I_{n_2} - A_{D2})^{-1} B_{D2} + \bar{D}_{r2}$$

$$R(\lambda) = \bar{C}_{r2}VT_D^{-1}T_c^{-1}\psi_r(\lambda) + \bar{D}_{r2}D_r(\lambda)$$

where $V = [0, I_{n_2}]$. From Lemma 5.2, we have

$$D_{rc2}(\lambda) = \bar{C}_{r2}(\lambda I_{n_2} - A_{D2} + B_{D2}F_{D2})^{-1}B_{D2}G_F + \bar{D}_{r2}G_F$$

Since $\hat{D}_{rc}(\lambda)$ is column reduced and $D_{rc2}(\lambda)$ is a canonical left divisor of $\hat{D}_{rc}(\lambda)$, from Theorem 4.10 and Eq. (6.15), we obtain

$$\hat{D}_{rc}(\lambda) = R'(\lambda)D_{rc2}(\lambda)$$

where

$$R'(\lambda) = \bar{C}_{r2}VT_D^{-1}T_c^{-1}\psi_r(\lambda) + \bar{D}_{r2}G_F\hat{D}_{rc}(\lambda)$$

Therefore

$$R'(\lambda) - R(\lambda) = \bar{D}_{r2}[G_F\hat{D}_{rc}(\lambda) - D_r(\lambda)]$$

Also, substituting Eq. (6.17c) into the above equation yields

$$R(\lambda) - R'(\lambda) = \bar{D}_{r2}F_D T_D^{-1}T_c^{-1}\psi_r(\lambda) = \bar{D}_{r2}[0, F_{D2}]T_D^{-1}T_c^{-1}\psi_r(\lambda)$$

Since $\bar{D}_{r2} = [I_m - \psi_{r2}^T(0)\psi_{r2}(0)]D_{rh2}^{-1}$, the ith row of F is null, and $(D_{rh2})_{ij} = (D_{rh2}^{-1})_{ij} = 0$ for $j \neq i$ if $\kappa_{2i} = 0$, it follows that

$$\bar{D}_{r2}F_{D2} = (I_m - \psi_{r2}^T(0)\psi_{r2}(0))D_{rh2}^{-1}F_{D2} \equiv 0$$

Therefore, $R(\lambda) = R'(\lambda)$, or $\hat{D}_{rc}(\lambda) = R(\lambda)D_{rc2}(\lambda)$.   ∎

From Eq. (6.17a) and Theorem 6.8 we obtain:

<u>Theorem 6.9</u>   The closed-loop RMFD of the modal controlled system can be represented as the semi-cascade form,

$$G_c(\lambda) = N_r(\lambda)R^{-1}(\lambda)D_{rc2}^{-1}(\lambda) \tag{6.20}$$

where $N_r(\lambda)$ is the numerator of the open-loop canonical RMFD, $R(\lambda)$ is the right divisor of the open-loop right characteristic λ-matrix associated with the uncontrolled modes, and $D_{rc2}(\lambda)$ is the desired closed-loop right characteristic λ-matrix of the controlled modes. ∎

Theorem 6.9 reveals the important feature of the modal control design that the left divisor of the denominator of the open-loop RMFD has been changed to the desired one with prespecified eigenstructures.

Note that $G_c(\lambda)$ in Eq. (6.20) may <u>not</u> be a canonical RMFD. Using Theorem 3.4, we can transform Eq. (6.20) into a canonical RMFD as follows:

$$G_c(\lambda) = N_{rc}(\lambda)D_{rc}^{-1}(\lambda) \tag{6.21}$$

where $N_{rc}(\lambda) = N_r(\lambda)U_r(\lambda)$, $D_{rc}(\lambda) = \hat{D}_{rc}(\lambda)U_r(\lambda)$, and $U_r(\lambda)$ is a unimodular λ-matrix. The result in Eq. (6.21) implies that although the zeros of the open-loop system remain unchanged under modal control laws, the numerator of the closed-loop canonical RMFD may be changed.

<u>Example 6.2</u>

Given a 3-input 2-output continuous time system in the form (6.2) where

$$A = \begin{bmatrix} 2.2 & 1.6 & 4.0 & 3.0 & 1.0 \\ -6.4 & -4.2 & -8.0 & -6.0 & -2.0 \\ -1.6 & -0.8 & -3.0 & -2.0 & -1.0 \\ 3.2 & 1.6 & 4.0 & 3.0 & 2.0 \\ 9.6 & 4.8 & 12.0 & 6.0 & 2.0 \end{bmatrix}$$

$$B = \begin{bmatrix} -2.0 & 1.0 & 0.0 \\ -1.0 & -2.0 & -5.0 \\ 3.0 & -1.0 & 1.0 \\ -2.0 & 3.0 & 4.0 \\ 1.0 & -2.0 & -3.0 \end{bmatrix}$$

$$C = \begin{bmatrix} 3.6 & 3.8 & 5.0 & 4.0 & 0.0 \\ 5.4 & 2.2 & 5.0 & 3.0 & -1.0 \end{bmatrix}, \qquad D = 0_{2 \times 3}$$

The RMFD of the system is found to be

$$G(\lambda) = N_r(\lambda) D_r^{-1}(\lambda)$$

where

$$N_r(\lambda) = \begin{bmatrix} -4\lambda^2 + 13\lambda - 13 & 7\lambda - 12 & 0.0 \\ -5\lambda^2 + 10\lambda + 3 & 12\lambda + 7 & 0.0 \end{bmatrix}$$

and

$$D_r(\lambda) = \begin{bmatrix} \lambda^3 - 2\lambda^2 - \lambda + 2 & -\lambda^2 + \lambda + 2 & -1 \\ 0 & \lambda^2 + 2\lambda + 1 & -2 \\ 0 & 0 & 1 \end{bmatrix}$$

We shall illustrate the semi-cascade realization and the modal control design for this system. For these purposes, the system map A has to be block-triangularized. We shall use Theorem 4.20 to construct the transformation T_D to transform A to a block-triangular form.

Selecting $r_0 = -\infty$ and $r_1 = 0$ and utilizing the matrix sign algorithm in Theorem 4.21 yields

$$\text{Sign}_{(r_0)}(A) = I_5$$

$$\text{Sign}_{(r_1)}(A) = \begin{bmatrix} -2.6 & -0.8 & -2.0 & 0.0 & 0.0 \\ 3.2 & 0.6 & 4.0 & 0.0 & 0.0 \\ 1.6 & 0.8 & 1.0 & 0.0 & 0.0 \\ -3.2 & -1.6 & -4.0 & -1.0 & 0.0 \\ 6.4 & 3.2 & 8.0 & 4.0 & 1.0 \end{bmatrix}$$

From Theorem 4.20, we have the canonical injection map $S_{(r_0,r_1)}$ and the canonical projection map $\bar{V}_{(r_0,r_1)}$:

$$S_{(r_0,r_1)} = \begin{bmatrix} 1.8 & -1.6 & -0.8 & 1.6 & -3.2 \\ 0.4 & 0.2 & -0.4 & 0.8 & -1.6 \\ 1.0 & -2.0 & 0.0 & 2.0 & -4.0 \end{bmatrix}^T$$

$$\bar{V}_{(r_0,r_1)} = \begin{bmatrix} -0.8 & -0.4 & -1.0 & 0.0 & 0.0 \\ 3.2 & 1.6 & 4.0 & 2.0 & 1.0 \end{bmatrix}$$

Thus the transformation T_D can be constructed as

$$T_D = \begin{bmatrix} S^+_{(r_0,r_1)} \\ \bar{V}_{(r_0,r_1)} \end{bmatrix}^{-1} = \begin{bmatrix} 0.6444 & -1.7778 & -0.4444 & -0.1000 & 0.2000 \\ -0.1778 & 0.9111 & -0.2222 & 0.2000 & -0.4000 \\ -0.4444 & -0.2222 & 0.4444 & 0.1000 & -0.2000 \\ -0.8000 & -0.4000 & -1.0000 & 0.0000 & 0.0000 \\ 3.2000 & 1.6000 & 4.0000 & 2.0000 & 1.0000 \end{bmatrix}^{-1}$$

$$= \begin{bmatrix} 1.8000 & 0.4000 & 1.0000 & -0.4444 & 0.0000 \\ -1.6000 & 0.2000 & -2.0000 & -0.2222 & 0.0000 \\ -0.8000 & -0.4000 & 0.0000 & -0.5556 & 0.0000 \\ 1.6000 & 0.8000 & 2.0000 & 1.6000 & 0.4000 \\ -3.2000 & -1.6000 & -4.0000 & 0.8000 & 0.2000 \end{bmatrix}$$

Defining

$$X(t) = T_D X_D(t)$$

then the state equations of the system in X_D - coordinates become

$$\lambda X_D(t) = A_D X_D(t) + B_D u(t)$$

$$y(t) = C_D X_D(t) + D_D u(t)$$

where

$$A_D = T_D^{-1} A T_D = \begin{bmatrix} A_{D1} & A_{D12} \\ 0_{2 \times 3} & A_{D2} \end{bmatrix}$$

$$B_D = T_D^{-1} B = \begin{bmatrix} B_{D1} \\ B_{D2} \end{bmatrix}$$

$$C_D = C T_D = [C_{D1}, \ C_{D2}]$$

and

$$A_{D1} = \begin{bmatrix} 0.6 & 0.8 & 2.0 \\ -3.2 & -2.6 & -4.0 \\ 0.0 & 0.0 & -1.0 \end{bmatrix} ; \quad A_{D12} = \begin{bmatrix} 2.0889 & 2.2444 \\ -1.9556 & -3.3778 \\ -0.4889 & -0.8444 \end{bmatrix}$$

$$A_{D2} = \begin{bmatrix} 1.0 & 1.0 \\ 0.0 & 2.0 \end{bmatrix}$$

$$B_{D1} = \begin{bmatrix} -2.0444 & 0.7444 & -0.5556 \\ -2.0222 & -0.3778 & -2.7778 \\ 2.0444 & 0.2556 & 2.5556 \end{bmatrix} ; \quad B_{D2} = \begin{bmatrix} -1.0 & 1.0 & 1.0 \\ 1.0 & 0.0 & 1.0 \end{bmatrix}$$

$$C_{D1} = \begin{bmatrix} 2.8 & 3.4 & 4.0 \\ 10.2 & 4.6 & 11.0 \end{bmatrix} ; \quad C_{D2} = \begin{bmatrix} 1.1778 & 1.6000 \\ -1.6667 & 1.0000 \end{bmatrix}$$

$$D_D = 0_{2 \times 3}$$

(1) Semi-Cascade Realization

From Corollary 4.2, $D_r(\lambda)$ can be factored as

$$D_r(\lambda) = D_{r2}(\lambda) R(\lambda)$$

where $D_{r2}(\lambda)$ is the right characteristic λ-matrix of (A_{D2}, B_{D2}):

$$D_{r2}(\lambda) = \begin{bmatrix} \lambda-2 & 0 & -1 \\ -2 & \lambda-1 & -2 \\ 0 & 0 & 1 \end{bmatrix}$$

and $R(\lambda)$, which is a right divisor of $D_r(\lambda)$, can be computed by using Eq. (6.11c):

$$R(\lambda) = \begin{bmatrix} \lambda^2-1 & -\lambda-1 & 0 \\ 2\lambda+2 & \lambda+1 & 0 \\ 0 & 0 & 1 \end{bmatrix}$$

Thus, from Lemma 6.1, the semi-cascade RMFD of the entire system is given by:

$$G(\lambda) = N_r(\lambda)R^{-1}(\lambda)D_{r2}^{-1}(\lambda)$$

Note that $D_{r2}(\lambda)$ contains two unstable modes of the system with latent roots 1 and 2, and $R(\lambda)$ contains three stable modes with latent roots -1,-1, and -1.

(2) Modal Control Law

Since A_{D2} contains two unstable modes with eigenvalues 1 and 2, we shall design a modal controller to stabilize these two modes. Using the left latent structure assignment in Theorem 5.4, we select the latent roots and latent vectors of $D_{rc2}(\lambda)$ in Theorem 6.7 to be

$$\bar{\lambda}_{21} = -2 \qquad\qquad P_{210} = \begin{bmatrix} 1 & -1 & 0 \end{bmatrix}^T$$

and

$$\bar{\lambda}_{22} = -4 \qquad\qquad P_{220} = \begin{bmatrix} 1 & 1 & 0 \end{bmatrix}^T$$

so we have

$$\bar{A}_{J2} = \begin{bmatrix} -2 & 0 \\ 0 & -4 \end{bmatrix}$$

and

$$\bar{B}_{J2} = \begin{bmatrix} 1 & -1 & 0 \\ 1 & 1 & 0 \end{bmatrix}$$

The right characteristic λ-matrix of $(\bar{A}_{J2}, \bar{B}_{J2})$ is found to be

$$D_{rc2}(\lambda) = \begin{bmatrix} \lambda+3 & 1 & 0 \\ 1 & \lambda+3 & 0 \\ 0 & 0 & 1 \end{bmatrix}$$

Thus, from Eq. (6.15a) and Theorem 6.7, we have the modal control law

$$u(t) = -FX(t) + G_F r(t)$$

where

$$F = F_D T_D^{-1} = - \begin{bmatrix} -18.4 & -9.2 & -23.0 & -12.0 & -6.0 \\ -19.2 & -9.6 & -24.0 & -14.0 & -7.0 \\ 0.0 & 0.0 & 0.0 & 0.0 & 0.0 \end{bmatrix}$$

and

$$G_F = \begin{bmatrix} 1.0 & 0.0 & -1.0 \\ 0.0 & 1.0 & -2.0 \\ 0.0 & 0.0 & 1.0 \end{bmatrix}$$

From Theorem 6.8, we have the RMFD of the closed-loop system:

$$G_c(\lambda) = N_r(\lambda) \hat{D}_{rc}^{-1}(\lambda)$$

where

$$\hat{D}_{rc}(\lambda) = D_{rc2}(\lambda) R(\lambda)$$

$$= \begin{bmatrix} \lambda^3+3\lambda^2+\lambda-1 & -\lambda^2-3\lambda-2 & 0 \\ 3\lambda^3+8\lambda+5 & \lambda^2+3\lambda+2 & 0 \\ 0 & 0 & 1 \end{bmatrix}$$

Note that $G_c(\lambda) = N_r(\lambda)\hat{D}_{rc}^{-1}(\lambda)$ is not a canonical RMFD. The canonical RMFD of the closed loop system can be found from the closed loop state equations:

$$\lambda X(t) = \bar{A}X(t)+BG_F r(t)$$

$$y(t) = CX(t)+Du(t)$$

where

$$\bar{A} = A-BF$$

$$= \begin{bmatrix} -19.8 & 10.4 & 26.0 & 13.0 & 6.0 \\ 50.4 & 24.2 & 63.0 & 34.0 & 18.0 \\ -37.6 & -18.8 & -48.0 & -24.0 & -12.0 \\ -17.6 & -8.8 & -22.0 & -15.0 & -7.0 \\ 29.6 & -14.8 & 37.0 & 22.0 & 10.0 \end{bmatrix}$$

From Eq. (2.15), we have the canonical RMFD of the closed-loop system:

$$G_c(\lambda) = N_{rc}(\lambda)D_{rc}^{-1}(\lambda)$$

where

$$N_{rc}(\lambda) = \begin{bmatrix} -4\lambda^2-8\lambda+23 & 7\lambda-12 & 0 \\ -5\lambda^2-26\lambda-18 & 12\lambda+7 & 0 \end{bmatrix}$$

$$D_{rc}(\lambda) = \begin{bmatrix} \lambda^3+6\lambda^2+10\lambda+5 & -\lambda^2-3\lambda-2 & 0 \\ -\lambda-1 & \lambda^2+3\lambda+2 & 0 \\ 0 & 0 & 1 \end{bmatrix}$$

It is easy to show that

$$D_{rc}(\lambda) = \hat{D}_{rc}(\lambda) U_r(\lambda)$$

and

$$N_{rc}(\lambda) = N_r(\lambda) U_r(\lambda)$$

where $U_r(\lambda)$ is a unimodular λ-matrix:

$$U_r(\lambda) = \begin{bmatrix} 1 & 0 & 0 \\ -3 & 1 & 0 \\ 0 & 0 & 1 \end{bmatrix}$$

Also, p_{210} and p_{220} are left latent vectors of $D_{rc}(\lambda)$.

6.3   Cascade Decomposition Theories and Their Applications to Multiport Network Synthesis

In this section, we shall present the factorization of MFDs to achieve the cascade realizations of MIMO systems. The minimal factorizations of MFDs are defined first. Algorithms to perform the minimal factorization [91-98] of RMFDs and LMFDs are derived. The corresponding state-space cascade realizations are also discussed. Finally, applications of the cascade system structures to multiport network realizations are illustrated.

6.3.1 Minimal Factorizations of MFDs

If $G(\lambda) \in C^{m \times m}[\lambda]$ can be factorized as

$$G(\lambda) = G_1(\lambda) G_2(\lambda) \tag{6.22}$$

where $G(\lambda)$, $G_1(\lambda)$ and $G_2(\lambda)$ are described by MFDs, then $G_1(s)$ and $G_2(\lambda)$ are called the left factor and right factor of $G(\lambda)$, respectively. The factorization of $G(\lambda)$ in Eq. (6.22) is minimal [94] if

$$\text{Deg}(D(\lambda)) = \text{Deg}(D_1(\lambda)) + \text{Deg}(D_2(\lambda)) \tag{6.23}$$

where $D(\lambda)$, $D_1(\lambda)$ and $D_2(\lambda)$ are the left or right characteristic λ-matrices of the systems described by $G(\lambda)$, $G_1(\lambda)$ and $G_2(\lambda)$, respectively. The minimal factorization of $G(\lambda)$ in Eq. (6.22) can be generalized as

$$G(\lambda) = G_1(\lambda)G_2(\lambda)\ldots G_k(\lambda) \tag{6.24}$$

with $\text{Deg}(D(\lambda)) = \sum_{i=1}^{k} \text{Deg}(D_i(\lambda))$.

Definition 6.4 An RMFD, $G(\lambda) = N_r(\lambda)D_r^{-1}(\lambda) \in C^{m \times m}(\lambda)$, is nonsingular if $D_r(\lambda) \in C^{m \times m}[\lambda]$ is a column-reduced canonical λ-matrix, and $N_r(\lambda) \in C^{m \times m}[\lambda]$ is nonsingular. Similarly, an LMFD, $G(\lambda) = D_\ell^{-1}(\lambda)N_\ell(\lambda)$, is nonsingular if $D_\ell(\lambda) \in C^{m \times m}[\lambda]$ is a row-reduced canonical λ-matrix, and $N_\ell(\lambda) \in C^{m \times m}[\lambda]$ is nonsingular. □

Theorem 6.10 (RMFD Factorization)

Consider a nonsingular canonical RMFD expressed as $G(\lambda) = N_r(\lambda)D_r(\lambda)$. Assume that $D_1(\lambda)$ and $\bar{N}_1(\lambda)$ are a column-reduced left canonical divisor and a row-reduced right canonical divisor of $D_r(\lambda)$ and $N_r(\lambda)$, respectively, or

$$D_r(\lambda) = D_1(\lambda)\bar{D}_2(\lambda) \tag{6.25a}$$

$$N_r(\lambda) = N_2(\lambda)\bar{N}_1(\lambda) \tag{6.25b}$$

Let

$$\bar{N}_1(\lambda)\bar{D}_2^{-1}(\lambda) = H_1(\lambda) + \hat{N}_1(\lambda)\bar{D}_2^{-1}(\lambda) \tag{6.25c}$$

where $H_1(\lambda)$ is a λ-matrix and $\hat{N}_1(\lambda)\bar{D}_2^{-1}(\lambda)$ is a proper RMFD.

Let (A_1, B_1, C_1, D_1) be a minimal realization of the RMFD, $\hat{N}_1(\lambda)\bar{\hat{D}}_2^{-1}(\lambda)$. Then,

$$G(\lambda) = G_2(\lambda)G_1(\lambda) \tag{6.26a}$$

where

$$G_1(\lambda) = N_1(\lambda)D_1^{-1}(\lambda) \tag{6.26b}$$

$$G_2(\lambda) = N_2(\lambda)D_2^{-1}(\lambda) \tag{6.26c}$$

and

$$N_1(\lambda) = \tilde{N}_1(\lambda) + D_2(\lambda)H_1(\lambda) \tag{6.26d}$$

$$D_2^{-1}(\lambda)\tilde{N}_1(\lambda) = \hat{N}_1(\lambda)\bar{\hat{D}}_2^{-1}(\lambda) \tag{6.26e}$$

$D_2^{-1}(\lambda)\tilde{N}_1(\lambda)$ is the LMFD of (A_1, B_1, C_1, D_1).

Proof:

Since

$$G(\lambda) = N_r(\lambda)D_r^{-1}(\lambda) = N_2(\lambda)\bar{N}_1(\lambda)\bar{\hat{D}}_2^{-1}(\lambda)D_1^{-1}(\lambda) = N_2(\lambda)[H_1(\lambda) + \hat{N}_1(\lambda)\bar{\hat{D}}_2^{-1}(\lambda)]D_1^{-1}(\lambda)$$

and (A_1, B_1, C_1, D_1) is a minimal realization of the RMFD, $\hat{N}_1(\lambda)\bar{\hat{D}}_2^{-1}(\lambda)$, we can determine the corresponding LMFD, using (A_1, B_1, C_1, D_1) as follows:

$$D_2^{-1}(\lambda)\tilde{N}_1(\lambda) = \tilde{N}_1(\lambda)\bar{\hat{D}}_2^{-1}(\lambda) \ .$$

Therefore

$$G(\lambda) = N_2(\lambda)D_2^{-1}(\lambda)[\tilde{N}_1(\lambda) + D_2(\lambda)H_1(\lambda)]D_1^{-1}(\lambda) = N_2(\lambda)D_2^{-1}(\lambda)N_1(\lambda)D_1^{-1}(\lambda)$$

and the result in Eq. (6.26a) follows. ∎

Theorem 6.10 provides a method for factoring a nonsingular canonical RMFD into the product of two nonsingular RMFDs. Since $D_2(\lambda)$ may not be in a column-reduced canonical form, $G_2(\lambda)$ may not be a canonical RMFD. However, using Theorem 3.4, we can convert $G_2(\lambda)$ into a canonical RMFD.

<u>Corollary 6.1</u> The RMFD factorization in Theorem 6.10 is minimal.

<u>Proof</u>:

$$\text{Deg}(D_1(\lambda))+\text{Deg}(D_2(\lambda)) = \text{Deg}(D_1(\lambda))+\text{Deg}(\bar{D}_2(\lambda)) = \text{Deg}(D_r(\lambda)) \quad\blacksquare$$

Since the left factor $G_1(\lambda)$ and right factor $G_2(\lambda)$ can be represented as nonsingular canonical RMFDs, we can repeatedly factorize $G(\lambda)$ to obtain

$$G(\lambda) = G_k(\lambda)G_{k-1}(\lambda)\ldots G_1(\lambda) \tag{6.27}$$

if the conditions in Theorem 6.10 are satisfied.

The main feature of the RMFD factorization in Theorem 6.10 is that the poles and zeros with associated latent vectors of $G_1(\lambda)$ and $G_2(\lambda)$ can be preselected by performing spectral factorizations of $N_r(\lambda)$ and $D_r(\lambda)$. This property is important in multiport network synthesis and factorization of a para-Hermitian MFD.

The dual results of the LMFD factorization are stated as follows.

<u>Theorem 6.11</u> (LMFD Factorization)

Given a nonsingular canonical LMFD as

$$G(\lambda) = D_\ell^{-1}(\lambda)N_\ell(\lambda) \tag{6.28a}$$

Let $D_\ell(\lambda) = \bar{D}_2(\lambda)D_1(\lambda)$ and $N_\ell(\lambda) = \bar{N}_1(\lambda)N_2(\lambda)$, where $D_1(\lambda)$ and $\bar{N}_1(\lambda)$ are a row-reduced right canonical divisor and a column-reduced left canonical divisor of

$D_{\ell}(\lambda)$ and $N_{\ell}(\lambda)$, respectively. Let $\bar{D}_2^{-1}(\lambda) \ \bar{N}_1(\lambda) = H_1(\lambda) + \bar{D}_2^{-1}(\lambda) \ \hat{N}_1(\lambda)$, where $H_1(\lambda)$ is a λ-matrix and $\bar{D}_2^{-1}(\lambda) \ \hat{N}_1(\lambda)$ is proper. Also, let (A_1, B_1, C_1, D_1) be a minimal realization of $\bar{D}_2^{-1}(\lambda) \times \hat{N}_1(\lambda)$. Then

$$G(\lambda) = G_1(\lambda) G_2(\lambda) \tag{6.28b}$$

where

$$G_1(\lambda) = D_1^{-1}(\lambda) N_1(\lambda); \ G_2(\lambda) = D_2^{-1}(\lambda) N_2(\lambda); \ N_1(\lambda) = H_1(\lambda) D_2(\lambda)$$

$$+ \ \tilde{N}_1(\lambda); \ \tilde{N}_1(\lambda) D_2^{-1}(\lambda) = \bar{D}_2^{-1}(\lambda) \hat{N}_1(\lambda)$$

$\tilde{N}_1(\lambda) D_2^{-1}(\lambda)$ is the RMFD of (A_1, B_1, C_1, D_1). ∎

Repeatedly applying Theorem 6.11, a nonsingular canonical LMFD can be factorized into a product of many nonsingular canonical LMFDs as

$$G(\lambda) = G_1(\lambda) G_2(\lambda) \ldots G_k(\lambda) \tag{6.29}$$

6.3.2 State-space Realization of the Cascade Systems

Let the state-space realizations of $G_1(\lambda)$ and $G_2(\lambda)$ in Eq. (6.26a) be

$$\lambda X_1(t) = A_1 X_1(t) + B_1 u_1(t) \tag{6.30a}$$

$$y_1(t) = C_1 X_1(t) + D_1 u_1(t) \tag{6.30b}$$

and

$$\lambda X_2(t) = A_2 X_2(t) + B_2 u_2(t) \tag{6.30c}$$

$$y_2(t) = C_2 X_2(t) + D_2 u_2(t) \tag{6.30d}$$

where A_i, B_i, C_i and D_i, $i=1,2$ are matrices of appropriate dimensions. For cascade connections, we let $u(t) = u_1(t)$, $y_1(t) = u_2(t)$, and $y(t) = y_2(t)$, where $u(t)$ and $y(t)$ are the input and output of the overall system, respectively.

Then, the state-space realization of the overall system becomes

$$\lambda \begin{bmatrix} X_1(t) \\ X_2(t) \end{bmatrix} = \begin{bmatrix} A_1 & 0 \\ B_2C_1 & A_2 \end{bmatrix} \begin{bmatrix} X_1(t) \\ X_2(t) \end{bmatrix} + \begin{bmatrix} B_1 \\ B_2D_1 \end{bmatrix} u(t) \qquad (6.31a)$$

$$y(t) = [D_2C_1, C_2] \begin{bmatrix} X_1(t) \\ X_2(t) \end{bmatrix} + D_2D_1u(t) \qquad (6.31b)$$

The above result can be generalized to the case of many subsystems as follows.

<u>Theorem 6.12</u>   Given a nonsingular MFD, $G(\lambda)$, which has a minimal factorization

$$G(\lambda) = G_k(\lambda)G_{k-1}(\lambda)\ldots G_1(\lambda) \qquad (6.32)$$

and (A_i, B_i, C_i, D_i) is the minimal realization quadruples of $G_i(\lambda)$, $i=1,2,\ldots,k$. Then, the state-space minimal realization quadruple of $G(\lambda)$, (A,B,C,D), can be expressed as

$$A = \begin{bmatrix} A_1 & 0 & \cdots & 0 & 0 \\ B_2C_1 & A_2 & \cdots & 0 & 0 \\ B_3D_2C_1 & B_3C_2 & \cdots & 0 & 0 \\ \cdot & \cdot & \cdots & \cdot & \cdot \\ B_{k-1}D_{k-2}\cdots D_2C_1 & B_{k-1}D_{k-2}\cdots D_3C_2 & \cdots & A_{k-1} & 0 \\ B_kD_{k-1}\cdots D_2C_1 & B_kD_{k-1}\cdots D_3C_2 & \cdots & B_kC_{k-1} & A_k \end{bmatrix}$$

$$B = [B_1^T, (B_2D_1)^T, (B_3D_2D_1)^T, \ldots, (B_{k-1}D_{k-2}\cdots D_1)^T, (B_kD_{k-1}\cdots D_1)^T]^T$$

$$C = [D_k\cdots D_2C_1, D_k\cdots D_3C_2, D_k\cdots D_4C_3, \ldots, D_kC_{k-1}, C_k]$$

$$D = [D_kD_{k-1}\cdots D_1]$$

Proof:

Theorem 6.12 can be proved by using induction on the number of cascade subsystems.   ∎

6.3.3 Cascade Realizations of Multiport Networks

If the matrix transfer function of a multiport network can be factorized as

$$G(\lambda) = G_1(\lambda)G_2(\lambda)\ldots G_k(\lambda)$$

the subnetworks $G_i(\lambda)$ can be constructed in a way such that each $G_i(\lambda)$ contains desirable latent roots and associated latent vectors of $G(\lambda)$. Then, we can realize each subnetwork with each $G_i(\lambda)$ and cascade them to construct the complete network for $G(\lambda)$. Since the factorization of $G(\lambda)$ is minimal, the realized network uses minimal active elements or reactive elements. The advantages of cascade structure were noted in the introduction to this chapter.

Example 6.3

Let the canonical right matrix fraction description of a 2-input 2-output network be

$$G(\lambda) = N_r(\lambda)D_r^{-1}(\lambda)$$

where

$$N_r(\lambda) = \begin{bmatrix} 1 & 5 \\ 0 & 1 \end{bmatrix} \begin{bmatrix} I_2\lambda + \begin{pmatrix} 5 & 0 \\ 10 & 10 \end{pmatrix} \end{bmatrix}$$

$$D_r(\lambda) = I_2\lambda^2 + \begin{bmatrix} 11.8 & -1.8 \\ 2.05 & 10.2 \end{bmatrix} \lambda + \begin{bmatrix} 8.75 & -10 \\ 2.05 & 9.2 \end{bmatrix}$$

The zeros of $G(\lambda)$ are -5, -10 and the poles of $G(\lambda)$ are $(-1,-1,-10\pm j1)$.

<u>Find</u> (1) The minimal realizations of D_r^{-1} in the controller canonical form and in the Jordan form.

(2) The minimal cascade factorizations of $G(\lambda)$.

(3) The cascade realization of $G(\lambda)$ in the state-space representation.

(4) The implementation of the cascade networks.

<u>Solution</u>

Using Lemma 2.1, the minimal realization of $D_r^{-1}(\lambda)$ in the controller canonical form can be found as

$$D_r^{-1}(\lambda) = \bar{C}_c(\lambda I_4 - A_c)^{-1}B_c + \bar{D}_c$$

where

$$A_c = \left[\begin{array}{cc|cc} 0.00 & 1.00 & 0.00 & 0.00 \\ -8.75 & -11.80 & 10.00 & 1.80 \\ \hline 0.00 & 0.00 & 0.00 & 1.00 \\ -2.05 & -2.05 & -9.20 & -10.20 \end{array}\right]$$

$$B_c = \left[\begin{array}{cc} 0 & 0 \\ 1 & 0 \\ \hline 0 & 0 \\ 0 & 1 \end{array}\right]$$

$$C_c = \left[\begin{array}{cc|cc} 1 & 0 & 0 & 0 \\ 0 & 0 & 1 & 0 \end{array}\right]$$

$$D_c = 0_{2\times2}$$

The corresponding Jordan form minimal realization of D_r^{-1} can be obtained by finding the Jordan forms minimal realization quadruple (A_J, B_J, C_J, D_J) from (A_c, B_c, C_c, D_c) as

$$D_r^{-1}(\lambda) = \bar{C}_J(\lambda I_4 - A_J)^{-1}B_J + \bar{D}_J$$

where

$$A_J = \left[\begin{array}{cc|cc} -1 & 1 & 0 & 0 \\ 0 & -1 & 0 & 0 \\ \hline 0 & 0 & -10 & 1 \\ 0 & 0 & -1 & -10 \end{array}\right]; \quad B_J = \left[\begin{array}{cc} 1 & 1 \\ 0 & 1 \\ \hline 1 & 0 \\ -1 & 1 \end{array}\right]; \quad \bar{C}_J^T = \left[\begin{array}{cc} 0.1 & -0.025 \\ -0.1 & 0.15 \\ -0.1 & -0.1 \\ 0.0 & -0.125 \end{array}\right]; \quad \bar{D}_J = 0_{2\times2}$$

Note that the latent roots of $D_r(\lambda)$ have been arranged in two groups; $\{-1,-1\}$ and $\{-10\pm j1\}$ as shown in the system map A_J. Using Theorems 4.9 and 4.12 together with the Jordan form minimal realization quadruple of $D_r^{-1}(\lambda)$, yields the spectral factorization of $D_r(\lambda)$ as follows:

$$D_r(\lambda) = D_L(\lambda)D_0 D_r(\lambda)$$

where

$$D_0 = I_2; \quad D_R(\lambda) = I_2\lambda + \begin{bmatrix} 0.8 & -0.8 \\ 0.05 & 1.2 \end{bmatrix}; \quad D_L(\lambda) = I_2\lambda + \begin{bmatrix} 11 & -1 \\ 2 & 9 \end{bmatrix}$$

In a similar manner, the spectral factorization of $N_r(\lambda)$ can be determined as

$$N_r(\lambda) = \begin{bmatrix} 1 & 5 \\ 0 & 1 \end{bmatrix} \left[I_2\lambda + \begin{pmatrix} 5 & 0 \\ 10 & 10 \end{pmatrix} \right]$$

Thus, $G(\lambda)$ can be represented as

$$G(\lambda) = N_r(\lambda)[D_L(\lambda)D_R(\lambda)]^{-1} = N_r(\lambda)D_R^{-1}(\lambda)D_L^{-1}(\lambda)$$

$$= \begin{bmatrix} 1 & 5 \\ 0 & 1 \end{bmatrix} \left[I_2\lambda + \begin{pmatrix} 5 & 0 \\ 10 & 10 \end{pmatrix} \right] \left[I_2\lambda + \begin{pmatrix} 0.8 & -0.8 \\ 0.05 & 1.2 \end{pmatrix} \right]^{-1} \left[I_2\lambda + \begin{pmatrix} 11 & -1 \\ 2 & 9 \end{pmatrix} \right]^{-1}$$

Applying Theorem 6.10 gives the cascade factorization of $G(\lambda)$ as

$$G(\lambda) = G_2(\lambda)G_1(\lambda)$$

where

$$G_2(\lambda) = \begin{bmatrix} 1 & 5 \\ 0 & 1 \end{bmatrix} \left[I_2\lambda + \begin{pmatrix} 1.8552 & -0.4414 \\ 1.6569 & 0.1448 \end{pmatrix} \right]^{-1}$$

$$G_1(\lambda) = \left[I_2\lambda + \begin{pmatrix} 6.0552 & 0.3586 \\ 11.6069 & 8.9448 \end{pmatrix} \right] \left[I_2\lambda + \begin{pmatrix} 11 & -1 \\ 2 & 9 \end{pmatrix} \right]^{-1}$$

The cascade realization of $G(\lambda)$, using Theorem 6.12, can be found as

$$G(\lambda) = G_2(\lambda)G_1(\lambda) = \hat{C}_r(\lambda I_4 - \hat{A}_r)^{-1}\hat{B}_r + \hat{D}_r$$

where

$$\hat{A}_r = \begin{bmatrix} -11.0 & 1.0 & 0.0 & 0.0 \\ -2.0 & -9.0 & 0.0 & 0.0 \\ -4.9450 & 1.3586 & -1.8552 & 0.4414 \\ 9.6069 & -0.0552 & -1.6569 & -0.1448 \end{bmatrix} ; \quad \hat{B}_r = \begin{bmatrix} 1 & 0 \\ 0 & 1 \\ 1 & 0 \\ 0 & 1 \end{bmatrix}$$

$$\hat{C}_r = \begin{bmatrix} 0 & 0 & | & 1 & 5 \\ 0 & 0 & | & 0 & 1 \end{bmatrix} ; \quad \hat{D}_r = 0_{2\times2}$$

The cascade realization of $G(\lambda)$ using four one-port R-C networks and summers but without using multiwinding transformers and integrators [84] is shown in Fig. 6.7. The poles and zeros of the first subnetwork, described by $G_1(\lambda)$, are $\{-10\pm j1\}$ and $\{-5,-10\}$, respectively whilst the poles and zeros of the second subnetwork, described by $G_2(\lambda)$, are $\{-1,-1\}$ and $\{\infty,\infty\}$, respectively. The realized cascade multi-port network contains the prespecified poles and zeros in each subnetwork.

199

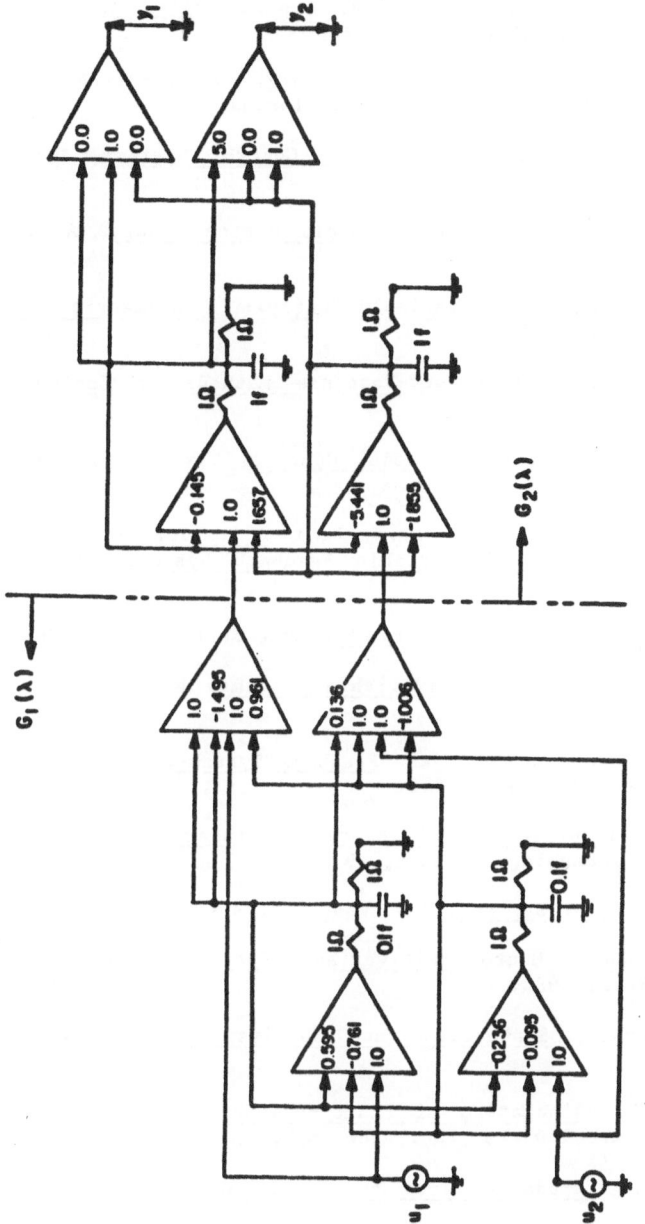

Fig. 6.7 Cascade Realization of the Network $G(\lambda)=G_1(\lambda)G_2(\lambda)$

BIBLIOGRAPHY

[1] R.E. Kalman, P.L. Falb, and M.A. Arbib, <u>Topics in Mathematical System Theory</u>, McGraw-Hill, New York, 1969.

[2] S. Barnett, <u>Matrices in Control Theory</u>, Van Nostrand Reinhold Co., New York, 1971; Second Edition, Krieger, Florida, 1984.

[3] S. Barnett, <u>Polynomials and Linear Control Systems</u>, Marcel Dekker, New York, 1983.

[4] M.K. Sain, <u>Introduction to Algebraic System Theory</u>, Academic Press, New York, 1981.

[5] W.M. Wonham, <u>Linear Multivariable Control: A Geometric Approach</u>, Springer Verlag, New York, 1979.

[6] A.G.J. MacFarlane, <u>Frequency-Response Methods in Control Systems</u>, IEEE Press, 1979.

[7] B. Porter and R. Crossley, <u>Modal Control</u>, Taylor & Francis Ltd., London, 1972.

[8] I. Postlethwaite and A.G.J. MacFarlane, <u>A Complex Variable Approach to the Analysis of Linear Multivariable Feedback Systems</u>, Springer-Verlag, New York, 1979.

[9] D.D. Siljak, <u>Large-Scale Dynamics Systems</u>, North-Holland, New York, 1978.

[10] W.A. Wolovich, <u>Linear Multivariable Systems</u>, New York: Springer-Verlag, 1974.

[11] C.A. Desoer and M. Vidyasagar, <u>Feedback Systems: Input-Output Properties</u>, Academic Press, New York, 1975.

[12] L. Pernebo, <u>Algebraic Control Theory for Linear Multivariable Systems</u>, Lund Institute of Technology, Sweden, 1978.

[13] T. Kailath, <u>Linear Systems</u>, Englewood Cliffs, NJ: Prentice-Hall, 1980.

[14] R.V. Patel and N. Munro, <u>Multivariable System Theory and Design</u>, Pergamon Press, New York, 1982.

[15] H.H. Rosenbrock, <u>State-Space and Multivariable Theory</u>, Nelson, London, 1970.

[16] D.L. Russell, <u>Mathematics of Finite-Dimensional Control Systems: Theory and Design</u>, Marcel Dekker Inc., New York, 1979.

[17] P. Lancaster, <u>Lambda Matrices and Vibrating Systems</u>, Pergamon Press, New York, 1966.

[18] J.E. Dennis, J.F. Traub, and R.P. Weber, "The Algebraic Theory of Matrix Polynomials", SIAM J. Numerical Analysis, 13, pp. 831-845, 1976.

[19] P. Lancaster, "A Fundamental Theorem on Lambda-Matrices with Applications-

I.   Ordinary Differential Equations with Constant Coefficients," Linear
Algebra and Its Applications, Vol. 18, pp. 189-211, 1977.

[20] I.   Gohberg,   P.   Lancaster,   L.   Rodman,   "Spectral   Analysis   of   Matrix
Polynomials I Canonical Forms and Divisors", Linear Algebra and Its
Applications, Vol. 20, pp. 1-44, 1978.

[21] I. Gohberg, M.A. Kaashoek, and L. Rodman, "Spectral Analysis of Families of
Operator Polynomial and a Generalized Vandermonde Matrox I, the Finite-
Dimensional Case", Academic Press, Topics in Functional Analysis, pp. 91-
128, 1978.

[22] I.   Gohberg,   P.   Lancaster,   and   L.   Rodman,   Matrix   Polynomials,   Academic
Press, New York, 1982.

[23] A.S. Markus and I.V. Mereuca, "On the Completen n-Tuple of Roots of the
Operator Equation Corresponding to a Polynomial Operator Bundle", Math.
USSR Izvestija, Vol. 7, pp. 1105-1128, 1978.

[24] L.S. Shieh, Y.T. Tsay, and N.P. Coleman, "Algorithms for Solvents and
Spectral Factors of Matrix Polynomials", Int. J. Systems Science, Vol. 12,
pp. 1303-1316, 1981.

[25] L.S. Shieh and Y.T. Tsay, "Transformations of Solvents and Spectral Factors
of Matrix Polynomials and their Applications", Int. J. Control, Vol. 34,
pp. 813-823, 1981.

[26] Y.T. Tsay, L.S. Shieh, R.E. Yates and S. Barnett, "Block Partial Fraction
Expansion of a Rational Matrix", Linear and Multilinear Algebra, Vol. 11,
pp. 225-241, 1982.

[27] L.S. Shieh and Y.T. Tsay, "Block Modal Matrices and their Applications to
Multivariable Control Systems", Proc. IEE, Vol. 129, Pt. D., pp. 41-48,
1982.

[28] L.S. Shieh and Y.T. Tsay, "Transformations of a Class of Multivariable
Control Systems to Block Companion Forms", IEEE Trans. Automat. Contr.,
Vol. AC-27, pp. 199-203, 1982.

[29] Y.T. Tsay and L.S. Shieh, "Some Applications of Rational Matrices to
Problems in Systems Theory", Int. J. Systems Science, Vol. 13, pp.   1319-
1337, 1982.

[30] Y.T. Tsay and L.S. Shieh, "Block Decomposition and Block Modal Control of
Multivariable Control Systems", Automatica, Vol. 19,   pp. 29-40, 1983.

[31] Y.T. Tsay and L.S. Shieh, "Irreducible Divisors of λ-Matrices and their
Applications to Multivariable Control Systems", Int. J. Control, Vol. 37,
pp. 17-36, 1983.

[32] L.S. Shieh, Y.T. Tsay and R.E. Yates, "Block Transformations of a Class of
Multivariable Control Systems to Four Basic Block Companion Forms", Int. J.
of Computers and Mathematics with Applications, Vol. 9, pp. 703-714, 1983.

[33] D.G. Luenberger, "Canonical Forms for Linear Multivariable Systems," IEEE
Trans. Automat. Control, Vol. AC-12, pp. 290-293, 1967.

[34] W.A. Wolovich and P.L. Falb, "On the Structure of Multivariable Systems",

SIAM J. Contr., Vol. 7, pp. 437-451, 1969.

[35] V.M. Popov, "Invariant Description of Linear Time-Invariant Controllable Systems", SIAM J. Contr., Vol. 10, pp. 252-264, 1972.

[36] R.E. Kalman, "Kronecker Invariant and Feedback", Ordinary Differential Equations, Edited by L. Weiss, Academic Press, pp. 459-471, 1974.

[37] A.C. Antoulas, "On Canonical Forms for Linear Constant Systems", Int. J. Control, Vol. 33, pp. 95-122, 1981.

[38] V. Kucera, Discrete Linear Control, John Wiley & Sons, New York, 1979.

[39] A.W. Naylor and G.R. Sell, Linear Operator Theory in Engineering and Science, Springer-Verlag, N.Y., 1982.

[40] L.S. Shieh and Y.T. Tsay, "A Minimal Nice Selection Algorithm for the Canonical Controller Structures of Multivariable Control System", U.S. Army Research Report, 1983.

[41] P. Lancaster and P.N. Weber, "Jordan Chains for Lambda Matrices", Linear Algebra and Its Applications, Vol. 1, pp. 563-569, 1968.

[42] F.P. Greenleaf, Introduction to Complex Variables, W.B. Saunders Co., Philadelphia, 1972.

[43] R.F. Rinehart, "The Equivalence of Definitions of a Matrix Function", American Mathematical Monthly, Vol. 62, pp. 395-414, 1955.

[44] L.S. Shieh and N. Chahin, "A Computer-Aided Method for the Factorization of Matrix Polynomials", Int. J. Systems Science, Vol. 12, pp. 305-323, 1981.

[45] J.E. Dennis, J.F. Traub and R.P. Weber, "Algorithms for Solvents of Matrix Polynomials", SIAM, Numer. Anal., Vol. 15, pp. 523-533, 1978.

[46] E.D. Denman, "Matrix Polynomials, Roots and Spectral Factors", Applied Mathematics and Computation, Vol. 3, pp. 359-368, 1977.

[47] P.P. Khargonekar and E. Emre, "Further Results on Polynomial Characterizations of (F,G)-Invariant and Reachability Subspaces", IEEE Trans. Automat. Control, Vol. AC-27, pp. 352-366, 1982.

[48] R.E. Kalman, "On Partial Realizations, Transfer Functions, and Canonical Forms", Act Polytechnica Scandinavia, pp. 9-32, 1979.

[49] R. Herman, Linear System Theory and Introductory Algebraic Geometry, Interdisciplinary Mathematics Vol. VIII, New Jersey, 1974.

[50] P.A. Fuhrmann, "Simulation of Linear Systems and Factorization of Matrix Polynomials", Int. J. Control, Vol. 28, pp. 689-705, 1978.

[51] P.A. Fuhrmann, "Algebraic System Theory: An Analyst's Point of View", Journal of the Franklin Institutes, Vol. 301, pp. 521-540, 1978.

[52] P.A. Fuhrmann and J.C. Willems, "A Study of (A,B)-Invariant Subspaces Via Polynomial Models", Int. J. Control, Vol. 31, pp. 467-494, 1980.

[53] A.C. Antoulas, "A System-Theoretic Approach to the Factorization Theory of

Non-Singular Polynomial Matrices", Int. J. Control, Vol. 33, pp. 1005-1026, 1981.

[54] M.S. Mahmoud and M.G. Singh, Large Scale Systems Modelling, Pergamon Press, N.Y., 1981.

[55] H. Kwakernaak and R. Sivan, Linear Optimal Control Systems, John Wiley & Sons, Inc. N.Y., 1972.

[56] J.D. Roberts, "Linear Model Reduction and Solution of the Algebraic Riccati Equation by Use of the Sign Function", CUED/B-CONTROL/TR 13 Report, Cambridge University, 1971, Also, Int. J. Control, 1980, Vol. 32, pp. 677-687.

[57] E.D. Denman and A.N. Beavers, "The Matrix Sign Function and Computations in Systems", Appl. Math. & Comput., Vol. 2, pp. 63-94, 1976.

[58] R.L. Mattheys, "Stability Analysis Via the Extended Matrix Sign Function", Proc. IEE, Vol. 125, pp. 241-243, 1978.

[59] L.A. Balzer, "Accelerated Convergence of the Matrix Sign Function Method of Solving Lyapunov, Riccati and Other Matrix Equations", Int. J. Control, 1980, Vol. 32, pp. 1057-1078.

[60] L.S. Shieh, Y.T. Tsay and R.E. Yates, "Some Properties of Matrix Sign Functions Derived from Continued Fractions", Proc. IEE, Vol. 130, Part D, pp. 111-118, 1983.

[61] C.T. Chen, Introduction to Linear System Theory, Holt, Rinehart and Winston, New York, 1970; Second Edition, 1984.

[62] A.J. Fossard, Multivariable System Control, North-Holland, 1972.

[63] B.S. Morgan, "Multivariable System Synthesis by State-Variable Feedback", IEEE Trans. Auto. Control, Vol. AC-9, pp. 405-411, 1964.

[64] B.C. Moore, "Eigenvalue and Eigenvector Assignment Using State Feedback", IEEE Trans. Auto. Control, Vol. AC-21, pp. 689-692, 1976.

[65] E.J. Davison, "Multivariable Systems Subject to Arbitrary Output and State Feedback", IEEE Trans. Auto. Control, Vol. AC-18, pp. 24-32, 1973.

[66] J.J. D'Azzo and C.H. Houpis, Linear Control System Analysis and Design, McGraw-Hill, New York, 1981.

[67] W.A. Wolovich, "The Differential Operator Approach to Linear System Analysis and Design", J. Franklin Inst., Vol. 301, pp. 27-47, 1976.

[68] W.A. Wolovich and P.L. Falb, "Invariants and Canonical Forms Under Dynamic Compensation", SIAM J. Contr., Vol. 14, pp. 996-1008, 1976.

[69] P.L. Falb and W.A. Wolovich, "The Roles of the Interactor in Decoupling", Proc. JACC, pp. 1456-1460, 1977.

[70] P.L. Falb and W.A. Wolovich, "Decoupling in the Design and Synthesis of Multivariable Control Systems", IEEE Trans. Auto. Control, Vol. AC-12, pp. 651-659, 1967.

[71] E.G. Gilbert, "The Decoupling of Multivariable Systems by State Feedback", SIAM J. Control, Vol. 7, pp. 50-63, 1969.

[72] H.H. Rosenbrock, "Design of Multivariable Control Systems Using Inverse Nyquist Array", Proc. IEE, Vol. 116, pp. 1929-1936, 1969.

[73] D.Q. Mayne, "The Design of Linear Multivariable Systems", Automatic, Vol. 9, pp. 201-207, 1973.

[74] D.H. Owens, "Dyadic Expansion, Characteristic Loci and Multivariable-Control System Design", Proc. IEE, Vol. 122, pp. 315-320, 1975.

[75] D.C. Youla, H.A. Jabr and J.J. Bongiorno, "Modern Wiener-Hopf Design of Optimal Controllers - Part II: The Multivariable Case", IEEE Trans. Autom. Control, Vol. AC-21, pp. 319-338, 1976.

[76] A.G.J. MacFarlane and I. Postlethwaite, "The Generalized Nyquist Stability Criterion and Multivariable Root Loci", Int. J. Control, Vol. 25, pp. 81-127, 1977.

[77] A.G.J. MacFarlane and I. Postlethwaite, "Characteristic Frequency Functions and Characteristic Gain Functions", Int. J. Control, Vol. 26, pp. 265-278, 1977.

[78] G. Bengtsson, "Output Regulation and Internal Models - A Frequency Domain Approach", Automatica, Vol. 13, pp. 333-345, 1977.

[79] B.W. Dickinson, T. Kailath and M. Morf, "Canonical Matrix Fraction and State-space Descriptions for Deterministic and Stochastic Linear Systems", IEEE Trans. Autom. Control, Vol. AC-19, pp. 656-667, 1974.

[80] J. Fadavi-Ardekani, S.K. Mitra and B.D.O. Anderson, "Extended State-space Model of Discrete-time Dynamical Systems. IEEE Trans. Circuits and Systems", Vol. CAS-29, pp. 547-556, 1982.

[81] J.C. Willems, "Almost Invariant Subspaces: An Approach to High Gain Feedback Design - Part I: Almost Controlled Invariant Subspaces", IEEE Trans. Autom. Control, Vol. AC-26, pp. 235-252, 1981.

[82] J.C. Willems, "Almost Invariant Subspaces: An Approach to High Gain Feedback Design - Part II: Almost Conditionally Invariant Subspaces", IEEE Trans. Autom. Control, Vol. AC-27, pp. 1071-1085, 1982.

[83] S.V. Rao and S.S. Lamba, "Suboptimal Control of Linear Systems Via Simplified Models of Chidambara", Proc. IEE, Vol. 121, pp. 879-881, 1974.

[84] L.S. Shieh, S. Yeh and H.Y. Zhang, "Realizations of Matrix Transfer Functions Using RC Ladders and Summers", Proc. IEE, Vol. 128, Part G, pp. 101-110, 1981.

[85] J. Tajima, H. Nagase and S.I. Takahashi, "Synthesis Problem of Voltage Transfer Function Matrix", Electronics and Communications in Japan, Vol. 61-A, pp. 17-25, 1978.

[86] F.F. Kuo, Network Analysis and Synthesis, John Wiley & Sons, New York, 1966.

[87] G.C. Temes and J.W. LaPatra, Introduction to Circuit Synthesis and Design,

205

McGraw-Hill, New York, 1977.

[88] R.W. Newcomb, Linear Multiport Synthesis, McGraw-Hill, New York, 1966.

[89] B. Porter, Synthesis of Dynamic Systems, Nelson, London, 1969.

[90] Y.T. Tsay and L.S. Shieh, "Block Modal Control Via Cascade Multivariable Control Structure", Int. J. Systems Science, Vol. 15, pp. 199-214, 1984.

[91] D.C. Youla, "On the Factorization of Rational Matrices", IRE Trans. Information Theory, Vol. IT-7, pp. 172-189, 1960.

[92] B.D.O. Anderson, "An Algebraic Solution to the Spectral Factorization Problems", IEEE Trans. Autom. Control, Vol. AC-12, pp. 410-414, 1967.

[93] B.D.O. Anderson, K.L. Hitz and N.D. Diem, "Recursive Algorithm for Spectral Factorization", IEEE Trans. Circuits and Systems, Vol. CAS-21, pp. 742-750, 1974.

[94] H. Bart, I. Gohberg and M.A. Kaashoek, Minimal Factorizations of Matrix and Operator Functions, Birhauser Verlag, Boston, 1979.

[95] H. Bart, I. Gohberg, M.A. Kaashoek and P. Van Dooren, "Factorizations of Transfer Functions", SIAM J. Control and Optimization, Vol. 18, pp. 675-696, 1980.

[96] P. Van Dooren and P. Dewilde, "Minal Cascade Factorization of Real and Complex Rational Transfer Matrices", IEEE Trans. Circuits and Systems, Vol. CAS-28, pp. 390-400, 1981.

[97] P. Dewilde and J. Vandewalle, "On the Factorization of a Nonsingular Rational Matrix", IEEE Trans. Circuits and System, Vol. CAS-22, pp. 637-645, 1975.

[98] L.S. Shieh, Y.T. Tsay and R.E. Yates, "Cascade Realization of Multiport Networks Via Factorization of Rational Matrix Functions", 7th Int. Symposium on the Mathematical Theory of Networks and Systems, Stockholm, Sweden, 1985.

[99] M.K. Sain, "The Growing Algebraic Presence in System Engineering: An Introduction", Proc. IEEE, Vol. 64, pp. 96-111, 1976.

[100] J.W. Helton, "Systems with Infinite-Dimensional State Space: The Hilbert Space Approach", Proc. IEEE, Vol. 64, pp. 145-160, 1976.

[101] J.L. Casti, "Recent Developments and Future Perspectives in Non-Linear System Theory", SIAM Review, Vol. 24, pp. 301-331, 1982.

[102] R.P. Roesser, "A Discrete State-Space Model for Linear Image Processing", IEEE Trans. Autom. Control, Vol. AC-20, pp. 1-10, 1975.

[103] B.G. Mertzios and P.N. Paraskevopoulos, "Transfer Function Matrix of 2-D Systems", IEEE Trans. Autom. Control, Vol. AC-26, pp. 722-725, 1981.

104] N.K. Bose, Multidimensional Systems: Theory and Applications, IEEE Press, New York, 1979.

[105] S.K. Mitra and M.P. Ekstrom, Two-Dimensional Digital Signal Processing, Dowden, Hutchinson and Ross, Stroudsburg, 1978.

[106] K.J. Astrom, <u>Introduction to Stochastic Control Theory</u>, Academic Press, New York, 1970.

[107] B.D.O. Anderson and J.B. Moore, <u>Optimal Filtering</u>, Prentice-Hall, Englewood Cliffs, New Jersey, 1979.

[108] G.J. Bierman, <u>Factorization Methods for Discrete Sequential Estimation</u>, Academic Press, New York, 1977.

[109] E.D. Denman, J. Leyva-Ramos and G.J. Jeon, "The Algebraic Theory of Latent Projectors in Lambda Matrices", Applied Mathematics and Computation, Vol. 9, pp. 273-300, 1981.

[110] L.S. Shieh, Y.T. Tsay and R.E. Yates, "State-Feedback Decomposition of Multivariable Systems Via Block-Pole Placement", IEEE Trans. Auto. Control, Vol. AC-28, pp. 850-852, 1983.

INDEX

Lecture Notes in Control and Information Sciences

Edited by M. Thoma and A. Wyner

Lecture Notes in Control and Information Sciences

Edited by M. Thoma and A. Wyner

Lecture Notes in Control and Information Sciences

Edited by M. Thoma and A. Wyner